STEAM TRA...

ANDREW MARTIN is a prolific author of both fiction and non-fiction, and his books often have a railway theme. His previous titles for Profile are *Underground Overground*, *Belles & Whistles* and *Night Trains*. His fiction includes the Jim Stringer series of historical crime novels set on early-twentieth-century railways. His website is *www.jimstringernovels.com*.

OTHER RAILWAY BOOKS BY ANDREW MARTIN

Seats of London

Night Trains

Belles & Whistles

Underground/Overground

FICTION

IN THE JIM STRINGER SERIES ...

The Necropolis Railway

The Blackpool Highflyer

The Lost Luggage Porter

Murder at Deviation Junction

Death on a Branch Line

The Last Train to Scarborough

The Somme Stations

The Baghdad Railway Club

Night Train to Jamalpur

Powder Smoke

STEAM TRAINS TODAY

Journeys Along Britain's Heritage Railways

ANDREW MARTIN

P

PROFILE BOOKS

This paperback edition published in 2022

First published in Great Britain in 2021 by
Profile Books Ltd
29 Cloth Fair
London
EC1A 7JQ
www.profilebooks.com

1 3 5 7 9 10 8 6 4 2

Typeset in Berling Nova Text by MacGuru Ltd
Printed and bound in Great Britain by
CPI Group (UK) Ltd, Croydon, CR0 4YY

A CIP catalogue record for this book is available
from the British Library.

ISBN 978 1 78816 145 9
eISBN 978 1 78283 489 2

'There is more to life than living in the present.'.

Keith Waterhouse, reviewing *The Rape of Britain* by
Colin Amery and Dan Cruickshank,
in the *Daily Mirror*, 12 June 1975

Contents

Some Terminology xi

Preface
Covid and the Heritage Lines xiii

Introduction
Mother's Day at Loughborough 1
The Swapmeet 1
Along the Line 12

1: Railway Preservation
Preserved or Heritage? 20
The Parallel Lines 25
Railway Preservation Before Beeching 30
Beeching Versus Betjeman 43

2: Some Pioneers
The Talyllyn Railway 53
The Booming of the Mountain Wind 53
Volunteer Platelayers Required 59
All Weathers and Conditions Bring Their Blessings 68
The Middleton Railway 73
The Hard Sell 77
The Bluebell Railway 84
Some Geography 84

The Sulky Service 92
Three Engines 98
Horsted Keynes 101
The Keighley & Worth Valley Railway 106
Cryer Meadows 106
Diversion to Barry 114
Along the Keighley & Worth Valley Railway 120
What Happened Next: Overview 126

3: Narrow-Gauge in England
The Gravitas of English Narrow-Gauge Railways 130
The Romney, Hythe & Dymchurch Railway 138
The Captains 138
Down the Rabbit Hole 144
Three Narrow-Gauge Seaside Lines in England 150
The Wells & Walsingham Light Railway 150
The North Bay Railway at Scarborough 155
The Lincolnshire Coast Light Railway 157

4: Narrow Gauge in Wales
The Ffestiniog Railway 159
Porthmadog 159
Observations 167
The Welsh Highland Railway 173
The Beyer-Garratt 173
Nutwood 175
The Other Welsh Highland Railway 178
The Great Little Trains of Wales 179
The Great Little Trains Differentiated 179
Rolt's Ghost and Jennie 184

5: Railways in the Landscape
Moss upon the Sleepers 188
Great Western Scenes 193
Didcot 193
The Swindon & Cricklade Railway 198
The West Somerset Railway 201
The Somerset & Dorset Joint Railway 205
'God's Own Country' 210
The North Yorkshire Moors Railway 210
The Wensleydale Railway 221
The Embsay & Bolton Abbey Railway 225
The Strathspey Railway 228
Problems in England 228
The Monarch of the Glen 231
By the Sea 236
The North Norfolk Railway 236
The Isle of Wight Steam Railway 242
The Swanage Railway 248

6: Specialists
Diesel 253
Diesel Romance 253
The Ecclesbourne Valley Railway 258
The Mid-Norfolk Railway 263
Diesel Gala on the Severn Valley Railway 268
Freight and Industrial Railways 273
The Romance of Freight and Industrial Railways 273
Rocks by Rail 280
The Leighton Buzzard Narrow-Guage Railway 284
The Watercress Line 289
The Kent & East Sussex: a Colonel Stephens Railway 295
The Colonel 295

The Eccentric Club 298
A Footnote: the Derwent Valley Light Railway 304
Foreign Trains in Cambridgeshire: The Nene Valley Railway 306
Special Days 310
A Car Day 310
Christmas Days 314
A Steam Driving Day 318
Thomas Days 324
Ale Train 333

7: Some Possible Futures
The Gloucestershire Warwickshire Railway 344
The East Lancashire Railway 351
Southwold 356

Select Bibliography 363
Picture Credits 365

Some Terminology

Railway Companies

In the Railway Grouping of 1923, more than a hundred railway companies were absorbed into four companies. These (the 'Big Four') were:

The Great Western (GWR)
The London, Midland & Scottish (LMS)
The London & North Eastern Railway (LNER)
The Southern Railway (SR)

Amongst the pre-Grouping companies mentioned in this book are:

The Great Western (the name persisted after the grouping, and indeed is used today by the privatised operator of the West Country routes)
The North Eastern Railway (NER)
The London & North Western Railway (LNWR)
The Midland Railway (MR)
The Lancashire & Yorkshire Railway (L&Y)
The Somerset & Dorset Railway (S&D)
The Highland Railway

The railways were nationalised in 1947, and British Railways came into being in 1948. It traded as British Rail from 1965.

In 1997, the railways were privatised.

Tank Engines and Tender Engines

A tank engine is a single unit: a steam locomotive carrying its coal in a bunker and its water in tanks close to its boiler. A tank engine is usually small. Bigger steam engines carry their coal and water in a trailer or tender.

Wheel formations

Under the Whyte system of notation, steam locomotives are described by their wheel formations. A 4–6–2, for instance, has four leading wheels, six driving wheels and two trailing wheels. There are always driving wheels, but there may not be leading wheels or trailing wheels, in which case a zero is used, as in 0–4–0. The driving wheels are coupled together: if there are, say, six of them, we might speak of a 'six-coupled engine'. Some of the wheel formations have names, usually with arcane origins: for example, 4–4–2 is an Atlantic, because it originated on the Atlantic Coast Railroad in America. There is also a system of describing the wheel formations of diesels, but we will not be concerning ourselves with that.

Preface

Covid and the Heritage Lines

When I was finishing this book in late 2019, I should have noticed that the title of the last chapter, 'Some Possible Futures', had a rather blasé ring about it. Readers will not find in that chapter any reference to a global pandemic, and in fact this book is entirely Covid-free, apart from this note.

Steam Trains Today was going to be published in March 2020, just when Covid hit. We decided to postpone publication for a year, because we wanted the heritage railways to be open when the book came out. Some sentences might read a bit oddly in light of the pandemic, especially those touching on the 'crowded commuter trains' of the national network. As for the heritage lines, a handful of the scenarios I describe have been overtaken by events. The Somerset & Dorset Railway Trust no longer has its museum at Washford, for instance, which is a shame. But for most of the year since the completion of this book, the lines have simply been engaged in a grim battle for survival. They were, like much of society, devastated by the pandemic, and many of the retirement-age volunteers who keep the lines going – 'Generation Steam', as they are known – will have sadly fallen victim to it.

In using that word 'devastated', I am quoting Steve Oates, Chief Executive of the Heritage Railway

Association, but Steve quickly added that, as of February 2021, 'The lines have all survived. I'm not aware of any that have closed.'

They were able to survive by furloughing paid staff; by generous public donations; by government money (from the Culture Recovery Fund and the Heritage Emergency Fund, among others) – and by their own resourcefulness. In 2020, Steve Oates himself had a holiday on the Talyllyn Railway in Wales, which, like many of the lines, opened in summer between lockdowns. In the interests of being Covid secure, the Talyllyn provided packed lunches to take on the trains (in the civilised tradition of the luncheon hampers available during steam days). 'It was just different and *nice*,' says Steve. Of that fleeting, flickering summer, he recalls, 'The lines that opened did all right, and some of them had their best September ever. There was a real hunger for them.'

At the time of writing, there seems a reasonable prospect that the lines will be open for at least some of the summer of 2021, and the 'hunger' could be even stronger, given that we might not be allowed to go abroad, and given that they permit just the kind of sociability and escapism we have long been denied. 'It could be a good summer,' says Steve Oates. 'It could be *brilliant*.'

I do hope so, and I would like to dedicate this book to all members, past and present, of Generation Steam.

Andrew Martin, *February 2021*

Introduction

Mother's Day at Loughborough

The Swapmeet

A rainy and dark Sunday morning in Leicestershire. I am standing on the edge of what appears to be a field of ash, and looking at a red-brick railway station, Quorn & Woodhouse by name. The station is not on the national network. You will not find it when booking on Trainline. It is a preserved station on a preserved line called the Great Central Railway, which is not to be confused with the *old* Great Central Railway (1897–1923), nine miles of whose 200-mile route the *preserved* Great Central has preserved. I will call these two the OGC and the PGC. The OGC route was closed between 1966 and 1969, on the recommendation of Dr Beeching, the BR chairman and railway 'axeman' who is going to be the villain of our piece. The PGC began operations in 1974, having bought its stretch of line from BR for £75,000.

We are already entering a historical thicket, as one tends to do on Britain's 120-or-so preserved railways, but a notice fixed to a fence between the ash field and the station provides some guidance. 'Opened in 1899 as part of the Great Central Railway's main line from Manchester and Sheffield to London Marylebone, the station is restored in the brown and "stone" colour scheme of the London & North Eastern Railway who owned the line in the 1940s.' So I probably ought to discount the 1970s diesel locomotive being rained on in a siding of the station, because that's strayed from a different time zone.

The preserved railways of Britain operate a time machine, and it does work, but it has a slight malfunction whereby it tends to default to the years of the Second World War. You're likely to be stepping out onto a station platform with – as here at Quorn & Woodhouse – posters reading 'Dig for Victory'. That standard wartime poster reading 'Is your journey really necessary?' is also here, even though it's a fatal question for the preserved railways, on which very few journeys are really necessary.

The PGC does re-create other periods at other stations, but let's stick with the 1940s for now. Many of the people involved in the preserved railways would have been born during the war, and a significant percentage will remember it; others might be connected to the 1940s by having been one of the quarter-million members of Ian Allan's Locospotters' Club, established in 1949 after Allan had been publishing his *ABC* Spotters' books for seven years. Many trainspotters gave up on spotting and became preservationists once British Rail had banished steam. I once asked Bob Gwynne, a curator at the National Railway Museum at York, and the author of a short book on preserved railways, whether the most favoured time-destination would creep forwards as the old guard died out.

He doubted it. 'The 1940s are a sort of fantasy or fairyland.'

'You mean like Narnia?' I suggested. 'Out of time?'

'Exactly.'

I do not think the preserved railways can be pinned

down politically, but somebody once suggested to me that they were 'a bit jingoistic'. The popularity of the wartime theme might reflect this, especially given it's an idealised war. In the real war, stations were blacked out, as were the trains, and the engines were run-down. All the festive aspects of railways – of which the preserved lines are so fond – were quelled. Trains ceased to have names, and in 1942 dining cars were withdrawn, so the PGC wouldn't have been allowed its Fish Supper Saturdays (always a sell-out). Perhaps one day a preserved line will have an authentic wartime event, with a soundtrack of bombers droning overhead, and everybody groping about in the half-light, as in A. A. Milne's verse:

We were alone, I hailed the fellow blindly,
'Excuse me, sir, I live at Wavertree.
Is it the next but one?' She answered kindly,
'It was the last but three.'

In Leicestershire, the rain and wind are increasing. Today is 11 March – Mothering Sunday, as all preserved railway operators are well aware. Mothering Sunday marks the start of their main season, which runs until October, although most have Christmas events, and some (like the PGC) operate all year round. There's an irony in the season kicking off with a women's day, since the majority of people who operate and appreciate the preserved railways are not women. At the Talyllyn Railway in Wales, I watched a man browsing in the bookshop while his wife looked on.

'And what are your own interests, madam?' the assistant, a young woman, asked the wife.

'Oh, I don't know,' she replied. 'I've been stuck with this railway lark for so long I've forgotten.' (She was joking, I think, and we will be meeting many women in this book.)

In a couple of hours' time, a Mothering Sunday Luncheon Special will be departing from Loughborough Central, one stop down the line. Meanwhile, another special event is taking place on the ash field behind me, which was once a goods siding: the annual PGC Swapmeet.

How much swapping is going on, I'm not sure. The event looks like a car boot sale of railway memorabilia, with trestle tables erected on the black cinder field, which is interspersed with oily puddles rippling in the wind – and by the way, it is PGC policy to evict anyone selling anything other than railway memorabilia, so there's no danger of the real world intruding. I examine a clothes brush stamped with the letters LBSCR: the London, Brighton & South Coast Railway, which ran fancy trains for fancy people, hence this clothes brush. The same stallholder goes in for bygone luxury, because he also offers some neatly pressed white antimacassars from the Blue Pullman. That was only a diesel train, and from the egalitarian 1960s, but it catered to wealthy businessmen, and was liveried not merely in blue but Nanking Blue.

Viewed from a modern perspective, the tone of the sale is peremptory, even politically incorrect. An armband says (underlinings are mine) 'Pilotman'. A small plastic sign reads, 'THIS LAVATORY MUST NOT BE FLUSHED WHILE THE TRAIN IS IN THE STATION'. A book on railway modelling has the subtitle, 'For the average enthusiast.' Some of the book titles are incredibly dour, chiming in with the greyness of the day: *From Tilbury to Tyneside:*

Great Eastern Railway Shipping; LMS Working Timetables: Supplement Number 4, Sectional Appendix.

A train – a green diesel multiple unit (DMU) – has pulled into the station. I have a rover ticket, so I can go up and down the line all day. The interior of the train is warm, an elegant pale green, with well-upholstered bench-like seats of grey-green. Everything throbs promisingly as the engines turn over, and there is a hot oil smell. The PGC has supplied more notices. First, a reproduction of an advert for this kind of train dating from the 1950s: 'Travel the Modern Way'. Well, diesel traction still is the modern way for a preserved railway, given that it postdates steam, and that there are as yet no preserved *electric* railways. Then there's a notice written by the PGC itself:

> Welcome on Board. You are travelling on a Diesel Railcar built in the 1950s. This style of train replaced steam power on many local and cross-country routes. At the time they were much welcomed because they were quicker, clean, well heated, had bright interiors and afforded excellent views along the line. Trains like this are known as diesel multiple units, as several can be joined together as needed.

It's a Class 101, one of a fleet made by Metro-Cammell of Birmingham. I like these because I remember them, whereas I have no childhood memory of steam, even though I was born in 1962, six years before its abolition. Therefore, I am not part of what the preserved railways call 'Generation Steam', on which they are dangerously reliant. The Class 101s I rode between Leeds and York were liveried in BR Rail Blue, a colour scheme introduced in 1965 by Misha Black. Blue was a very Sixties colour, fittingly unnatural in that futuristic decade of plastic and man-made fibre. On its carriages, BR combined the blue with off-white, hence the nickname 'blue and dirt'. BR Blue came too late for steam engines, but new diesels were painted this colour, and the old green ones were *re*painted in it. This present DMU that I'm sitting in must have then been re-painted *again*, back to green. An attractive characteristic of these early DMUs was that there was no interior bulkhead, but only a glass window between the driver's cab and the front passenger seats. As a boy I would always sit at the front, so as to share the driver's view of the track ahead ... except that some grumpy – probably hung-over – drivers, stumping on board at York with their flask of tea and riveted leather BR bag, would immediately turn around and lower the blinds over the windows between them and me, spitefully removing the forward view.

This being an idealised railway, the driver has *not* lowered the blind. He is a human tourist attraction and wants you to watch him driving. Moreover, the front seats here are First Class – dark blue with white antimacassars – and there's no extra charge for First on this preserved line. (There is on some.) The effect is of sitting in a luxury cinema showing a film of the line as it stretches away to Loughborough. You might say it's a Cinemascope film because there is also the side view. In one lineside field, a couple of people in long macintoshes are doing some serious dog training. This is hunting country, after all. The railway embankments are rather overgrown, but that's deliberate: it helps the wildlife and nesting birds.

The Class 101s on the PGC are owned by a sub-group of the railway called Renaissance Railcars. The diesels or steam engines on a preserved line might be owned directly by the line, but very often they're owned by separate syndicates of enthusiasts, like racehorses. (I mean to take a laissez-faire approach to the administrative organisation of the preserved railways, which quickly becomes complicated in most cases.) Class 101s finally disappeared from the real world in 2003, i.e. quite recently, so there are many on the preserved railways, for example at the Spa Valley, the Mid-Norfolk, the North Norfolk, the North York Moors Railway, the Foxfield and Ecclesbourne Valley Railways.

My feet, prone to be cold on modern, air-conditioned trains, are nice and warm, thanks to hot air emerging from a mysterious (to me) silvery duct next to my left boot. At Loughborough Central Station I alight from the train alongside the guard, who's about seventeen, and his vintage BR uniform, possibly bought last year at the swapmeet I've just attended, is too big for him. I decide to probe the mystery of the silver duct, reasoning that this will be a win-win. I'll increase my railway knowledge – and I do always feel I ought to be learning things on the preserved lines, perhaps to atone for the sheer pleasure I get from simply sitting on old trains – and the kid will presumably be pleased to enlighten me. 'How is the heat generated?' I ask. 'It's not steam heat, is it?'

'Oh, it's electric heating, I think,' he says, and he heads off to the buffet, but he's checked by the driver of the DMU, who has stepped onto the platform. The drivers on preserved lines tend to be in their sixties or seventies – seventy-five is the cut off – but this driver is a long-haired, scruffy-yet-capable-looking bloke perhaps in his forties.

'A *word*,' he says beckoning the kid. 'These trains are *not* electrically heated.' He then gives the kid a short lecture about how 'red diesel' is atomised in a cylinder, the resulting hot air being pushed by a fan into the heating ducts. The lecture concluded, the boy hurries off, slightly embarrassed.

I feel a bit guilty: the boy's a volunteer, so it's not as if he'd failed to do his job, but the preserved lines are two things: tourist attractions and living museums, and some of the volunteers are more curatorial than others. At least, I assume the driver's a volunteer, but some of the preserved railways do employ drivers, and so perhaps professional pride had prompted this one to correct the kid.

A professional driver of steam engines or old diesels is doing what many people would love to do for nothing – so the pay's terrible: 'About fifteen grand a year', I was told. Catering and management staff might also be paid. The PGC has 500 active volunteers, and about fifty paid staff, unusually high numbers in both cases, but then it is open all year round, and runs regular dining trains. A preserved line might be considered successful if it has the ratio of paid staff to volunteers that it *wants* to have: a high number of paid staff might indicate commercial and operational ambition, or it could reflect poor volunteer recruitment.

Loughborough Central features in Simon Jenkins' *Britain's 100 Best Railway Stations*, described as 'the archetypal railway museum'. The station has an early 1950s theme, hence pictures of the newly crowned queen. There's a big perambulator with a doll inside and a notice reading, 'Would station staff please return this pram to the waiting room at the end of the day,' because this is a theatrical prop, and preserved railways are a kind of amateur dramatics. The station is painted dark blue and green. A useful book to carry when visiting any preserved line would be *Station Colours* by Peter Smith. Having read it myself, I can offer some insight into why preserved railway people have to think hard before ordering their paint.

The stations on the OGC eventually became green, having been other colours earlier on. At the Railway Grouping of 1923, the OGC was taken into the London & North Eastern Railway. At first, the LNER painted its stations dark brown and stone (pale brown), and Quorn & Woodhouse on the PGC is in those colours, having a late 1930s/early 1940s theme. Later, a green and cream livery was introduced on LNER stations (and you can see that at, say, Pickering on the North Yorkshire Moors Railway). Loughborough Central, as mentioned, has a 1950s theme, by which time the LNER had been absorbed by the Eastern Region of British Railways. The different regions of BR had their different colours. For the Eastern Region, this was dark blue, but this was only used for signs and poster boards and other minor fittings. It was usually combined with the green that was left over from LNER days, and I think this is why Loughborough Central is green with blue signs. In other words, they've got it right.

Loughborough Central has a lovely, airy gentlemen's toilet with a high glass roof, white tiles and an aspidistra on a table. The best Gents in the UK are on the preserved railways, and I imagine the Ladies are pretty good too. It's possible that, when originally built in the late 1890s, this loo had no roof at all. You often find the roofless kind on preserved lines, and they're fine by me: the breeze removes the smell, and the rain washes the urinals.

The station is busy and bustling; the melancholia that can attend a wet Sunday afternoon is banished by signs informing me that the Station Master is 'in', the museum 'open'; the Refreshment Room and Fully Licensed Bar is also 'open' and a real fire burns within, as it does in the Ladies' Waiting Room and the General Waiting Room. Nearby is the engine shed, and there are 'Shed Tours every 55 minutes'. There is a bowl of clean water 'for dogs'. And a picturesque train waits at the platform, with carriages coloured umber and cream, the effete livery of the Pullman company, which operated luxury dining carriages in Britain until the early 1970s.

'Is it a real Pullman?' I ask a livered volunteer.

'Er, no,' he says ruefully, and I admire his directness, but a better answer would have been, 'It is, and it isn't.' The carriages are ordinary BR stock from the 1960s, mocked up as dining Pullmans in 2010. But the PGC does have permission to use the Pullman name. I peer inside: silver service, flowers in slender vases. Mothering Sunday Luncheon is about to be served, with a main course of roast Gressingham Duck Breast with Morello Cherry sauce, and a complimentary rose for Mother.

A big black steam engine, numbered 92214, is simmering away at the front. The driver stands, proudly proprietorial, alongside. A queue of people (all right, a queue of *men*) has formed, waiting to ask him a question, and I feel obliged to join it.

Now, the important thing is not to ask a *stupid* question. I have a theory about the engine – that it is a thing called a 9F* – but in the event I daren't risk the speculation, so I just ask, 'What is it?'

'A 9F,' says the driver.

'I *thought* so,' I reply, and the driver is polite enough not to appear sceptical.

This one at Loughborough is called *Leicester City* – as in Leicester City FC, those local heroes, and the name was given in honour of their winning the Premier League title in 2015–16. Only one of the 9Fs was given a name *at the time of building*: number 92220, and it was called *Evening Star*, in honour of being the last steam locomotive built by BR (in March 1960). It was always intended that it would be preserved: a rare bit of foresight in the flurried and indecisive period.

To compound the fatalism or surrealism surrounding *Evening Star*, it was the 999th Standard Class locomotive to be built, and Eric Treacy (the Bishop of Wakefield, known, on account of his railway enthusiasm, as 'the Railway Bishop') died of a heart attack while waiting to photograph it at Appleby Station on the Settle & Carlisle Line in 1978. I suppose it has often been said, 'It's how he would have wanted to go.' Today *Evening Star* is in the National Railway Museum at York, along with 12,000 of Eric Treacy's railway photographs.

Number 92214 is enjoying a more active retirement, drawing genteel diners up and down the PGC, while never exceeding 25 mph and pausing on the Swithland viaduct so passengers can admire the views of the Charnwood

* 9Fs were the biggest of the 'Standard' locos built by BR – locos built to BR's own design, rather than following the templates inherited from the 'Big Four'.

forest and reservoir – a bit of a come-down from hauling gigantic coal trains through the South Wales valleys, but it can consider itself lucky to be doing anything at all, having mouldered for fifteen years in the Woodham Brothers scrapyard at Barry, of which hellish spot we will be hearing more.

Along the Line

Over on Platform Two, another steam train has pulled in. Its engine is mid-sized and called *Witherslack Hall*. Odd, you might think, to name an engine after such an un-aerodynamic thing as a large country house, but then the Great Western Railway, which built this engine, also named engines after castles. 'Where were these historic "Halls", many of which one couldn't even pronounce?' asked Colin Garratt in *British Steam Nostalgia*; 'what was their history? Who lived in them? Yes, they came from a dream world which, having been glimpsed, heightened our childhood sense of wonderment and insight into the monumentality of history and creation ...' We are talking about a time (the inter-war period) when the railways wanted to appear as English and patrician as possible, in order to imply that the road hauliers undermining their business were somehow unpatriotic or insubordinate.

In fact, *Witherslack Hall* is a Modified Hall, a second generation of the original class. Other modified Halls are at the Swindon Railway Centre and West Somerset Railway. 'Modified Hall' is a cumbersome name, but this engine has the characteristic Great Western Railway elegance. The dark green – Brunswick Green – makes the lighter green of the station seem slightly parvenu; the chimney has the gleaming copper cap of the GWR, and the boiler has the

GWR characteristic of pointing slightly downwards, angled in relation to the tender as the barrel of a rifle is to the stock, and looking just as purposeful. Its driver is explaining some or all of the above as he conducts a small seminar on the footplate, while his colleague the fireman poses for pictures. This engine, too, is a Barry survivor.

The Halls, original and Modified, were a numerous class, and might well have turned up on the OGC. The line was noted for the variety of its traffic, ranging from luxury expresses – some hauled by *Flying Scotsman*, which was based at Leicester Central for a while – down to incessant trains of southbound coal.

The rake of carriages attached to *Witherslack Hall* is liveried in crimson lake and cream, an early BR colour scheme that was nicknamed blood and custard, and is not to be confused with the extremely similar 'plum and spilt milk', which was basically a London & North Western Railway colour scheme, although it too was briefly used on BR carriages. The carriages are BR Mark 1s, which is the commonest type of carriage found on the preserved railways. The Mark 1s were produced in the 1950s and 1960s and included both open seating and compartments. Like steam engines they are vacuum-braked, which means the two are easily compatible. You can see a Mark 1 interior in Christopher Nolan's 2017 film, *Dunkirk*. Harry Styles and his mates, returning from the Evacuation, are sitting on one of the characteristic Mark 1 bench-type seats with a small wooden wing at head height ... which is completely wrong, since the film is set in 1940 and Mark 1s date from the 1950s.

I think the seat covering – the moquette – on these Mark 1s *is* authentic, and it is in a variety of colours and patterns, all more exuberant than anything you'll see on modern trains. My favourite on this train is a sort of electric blue combined with yellow shapes suggesting autumnal leaves – like something dreamed up by William Morris. This moquette appears in a first-class compartment of six seats, two of which will prove to be occupied throughout the day by a couple slumbering so deeply that I am deterred from entering. In the same carriage are some second-class compartments (also six seats) in a warm, pinky red, and you have to look very hard to see the difference in the levels of luxury between First and Second. The second-class seats are slightly less plush, but both compartments offer what are in effect armchairs.

On this train there are also 'open' carriages (no compartments), and these are all Second Class, being built before 1987, when Second Class became 'Standard'. They are slightly smaller than the second-class compartment seats, but still wider than modern Firsts. In those days, carriage designers weren't trying to cram as many people in, which is why all seats are at a table and all are aligned to a window, a grace note absent even from the new Eurostar trains. The cosiness is enhanced by 30-watt bulbs and, as in a good pub, wood panelling. The earliest Mark Is featured a small plaque telling you what type of wood had been used. Here it's 'Crown Elm Britain'.

You could – and can – open the windows; or at least there is an openable portion, with sliding panes, and arrows to indicate the points beyond which a draught will be created. But sometimes you want a draught. In railway slang, modern air-conditioned carriages, with unopenable windows, are 'coffins'. Most surviving Mark Is do not meet modern crash regulations, so these days are only suitable for the preserved railways with their 25 mph top speeds, which is a shame, because their fitness for purpose is indicated by how frequently I fall asleep in Mark 1s, lulled by the tambourine jangle of the train, the steam heat fug and the lack of announcements. (There are no tannoys on Mark 1s.)

There are treats in store when you make little excursions from your seat. The WCs are roomy and pale yellow, with an appropriately lemony smell emanating from the cake of pungent soap. The sink is big enough to wash your face in, and there's a wide mirror to check the result. Everything is elemental and un-fiddly. The flush is a simple lever. You press down the taps, which you could do with your elbow, like a nurse.

There's also a buffet car, as there almost always is on a preserved line and usually not on the trains of the national network. On this train, hot food can be ordered throughout the day, and is brought to your seat wherever you happen to be sitting. But the main thing is the booze. Buffets on preserved trains are an alcoholic's dream (or nightmare), the shelves behind the server being stuffed with upended spirit bottles, real ales and ciders.

Our train has just gone through Quorn & Woodhouse, where the swapmeet was being packed up. We are now in deep country. Little ghosts of steam, bulletins from the engine, are spinning past in the rainy haze, and it occurs to me that the glamour of a steam locomotive shares something with that of an aeroplane: a heavy machine moving through water vapour. We are approaching Rothley Station, which has an Edwardian theme (i.e. late OGC, therefore green). The gas lamps are not yet lit, but they will be shortly.

Soon – too soon – we've hit the southern terminus of the PGC, Leicester North. Alone of the stations on the PGC, Leicester North has been entirely rebuilt. After the closure by Beeching it was demolished – not officially, but by vandals, which is almost impressive when you consider the solidity of the original. Beyond the buffers lies a railing, a small car park, and then the ghost railway – the old Great Central line to London Marylebone.

The OGC (the last main line built in Britain) was the brainchild of the flamboyant Sir Edward Watkin, who also ran the Metropolitan Railway. He intended the line to connect up with the Metropolitan in London, then to head south, delving into a Channel Tunnel and making for Paris, where Watkin kept at least one mistress. The line was built to accommodate European trains, which were, and are, wider and taller than English ones. This partly explains why the smaller stations along both the OGC and the PGC have island platforms, with tracks on either side – so that the trains would have elbow room. The PGC is the only preserved railway to have taken over a piece of main line; most run along former country branches. It even has a doubled section – two tracks side by side – which is unusual on a preserved line, and permits the drama (on the right running day) of one steam train passing another, and the Great Central is licensed, under certain conditions, to run trains at 45 mph rather than the 25 mph to which most preserved line trains are restricted.

Actually, the line beyond Leicester North is not ghostly all the way. The stretch between Aylesbury and Marylebone survives, and is used by a real-world company, Chiltern Trains. The rest was closed in stages between 1966 and 1969, thanks to Dr Beeching, who regarded it as a duplicate of the Midland Main Line. It's the only English main line he closed, and if he *hadn't* closed it, we might not be having to spending £80bn (or whatever is the latest estimate) building High Speed 2, because the Great Central Line, with its wide clearances, would have suited high-speed trains.

The PGC has, like most preserved railways, 'a project': a subject to which discussion turns as the fires are being dropped from the engines at the end of the day, or in the pub afterwards. Insofar as these projects are discussed more formally, I imagine a Captain Mainwaring-like figure pointing with a swagger stick at a complicated diagram on a blackboard and saying things like, 'I anticipate that a number of different volunteers will be required as the working party progresses.'

In the case of the PGC, the diagram would show the opposite end of the line to Leicester North: Loughborough Central, from where the PGC wants to probe north, connecting up with another preserved railway called the Great Central Railway (Nottingham) that has been formed on another stretch of OGC. The Great Central Railway (Nottingham) is an outgrowth of the Nottingham Transport Heritage Centre at Ruddington Fields (I hope readers are beginning to get an idea of the density of railway preservation in Britain). The Great Central (Nottingham) is a smaller concern than the PGC, with fewer running days and only two stations, but it might be said to have more gravitas, in that it is not merely a tourist line: it has a link to the national network which is used by freight trains hauling loads of gypsum.

With its 9 miles, the PGC is of about average length for a preserved railway. When it connects with its neighbour to the north, the joint concern will have 16 miles. (The preserved Middleton Railway in Leeds is less than a mile long, whilst the West Somerset is the longest standard-gauge preserved line with 22½ miles. The Welsh Highland and Ffestiniog, which connect and are run as a single enterprise, have a combined length of 34 miles, but they are narrow-gauge.)

For the two railways to reunite, a gap – known in Loughborough, where the lacuna occurs, as *The* Gap – must be bridged. The Gap developed in the 1970s, after the closure of the OGC. The bridge that carried the OGC over the Midland Main Line was demolished, and another bridge carrying the OGC across the Grand Union Canal fell into disrepair. The bridge over the Midland Main Line has now been replaced, with funding from the Leicestershire Local Enterprise Partnership and a public appeal. The canal bridge is being refurbished, but there is yet another bridge to be built (traversing a street), plus a length of embankment. In short, we must banish the image of bumbling Captain Mainwaring. This is a multi-million-pound project, one of the biggest undertaken by any preserved railway, and here is one of the paradoxes of preservation: on the one hand is the whimsy – the dolls in old prams on station platforms; on the other hand the feats of heavy engineering. The PGC speaks of the new bridge over the Midland Main Line as a 'bridge to the future', but it has been funded, in essence, by the romantic attraction of steam trains, a phenomenon that might or might not outlive 'Generation Steam'.

Back at Loughborough Central, the dining train has departed, but the place is still busy, with everything still open and the fires still burning. The exit from the station is via the wood-panelled, gaslit booking office, where slightly confused people sometimes ask for a ticket to somewhere as far afield as Rugby or even, in extreme distraction, London. (They are politely directed to Loughborough's real-world station about a mile away.) I walk through the booking office and step into the street, where it's cold. An ugly 4x4 is reversing in the rain. When it drives off, I am left with the view of an automotive repair shop. I turn on my heel, and head quickly back inside to the warm steam fug, the coal fires and the early 1950s.

1
Railway Preservation

Preserved or Heritage?

According to the latest figures from the Heritage Railway Association (HRA) there are about 180 'heritage railway attractions' in Britain, running over 500 miles of track – more than the London Underground network – and the figure is growing all the time. At the time of writing, they employ nearly 4,000 people and benefit from the work of 22,000 volunteers. They carry 13 million people a year and contribute half a billion pounds a year to the British economy. About 120 of those attractions could be called preserved railways per se, but it's hard to be precise: are we to include tramways (I don't intend to), and what about static exhibitions in museums? (I will be mentioning some of those).

We ought also to decide between 'preserved' and 'heritage' railways. Or perhaps there's nothing to choose between them. As we will be seeing, the forerunners of the present-day Railway Heritage Association were the Association of Railway *Preservation* Societies and a Railway *Preservation* Association. Both terms are problematic. When I visited the Romney, Hythe & Dymchurch Railway – a narrow-gauge steam railway line in

Kent – one volunteer denied it was a 'preserved railway', and with some ferocity.

'How *can* it be? We never closed. We've been running continuously since 1927.' He rejected the term 'heritage railway' with equal force. (I should say that I had irritated this chap by questioning the RH&D's claim to be 'the smallest public railway in the world'.) 'We're not heritage *or* preserved,' he trenchantly summarised. 'We're just the Romney, Hythe & Dymchurch Railway. End of.'

But imagine that you are lucky – or unlucky – enough to live in the stony wilderness of Dungeness, where the RH&D terminates next to the nuclear power station. You are giving instructions to a person unfamiliar with the locale who is proposing to visit you by public transport. I don't know how many buses go to Dungeness, but the only train that does (excluding the one that takes nuclear waste away from the power station) is the RH&D. So would you go into further details about the line, possibly bringing in words like 'heritage' and 'preserved'? Or would you just leave your visitor to discover a train of tiny carriages attached to a highly polished steam engine that is not as tall as they are? In which case, they might be wondering whether they've found the right line or drifted off into a dream world.

There are other railways that seem ambiguously poised between modern-day service and museum. The Island Line on the Isle of Wight, which is not to be confused with the preserved Isle of Wight Steam Railway, uses former Tube trains that date from 1938, so its trains are older than many on preserved railways. Islands off the British mainland tend to have endemic trains just as the Galapagos Islands have endemic fauna. The Isle of Man

Railway is a narrow-gauge steam railway that has been running almost continuously since 1873 (when it would not have looked particularly quaint) and is today operated by the Isle of Man government. For a couple of years in the late 1960s, no service ran because the money ran out; but if we class the resumed service as either 'preserved' or 'heritage' we might be tempted to apply the same term to any line that reopens after any closure.

The Settle & Carlisle railway also looks like a museum piece, even though it's on the national network. As with many of the preserved railways, Dr Beeching recommended its closure, and its future had to be fought for. It generally seems too beautiful for the modern world, and steam trains frequently run along it. (Steam locos on the national network are attended by a special team of experts, sometimes following in cars, and are watered from tanker lorries, since there are no longer any water troughs on the national network.)

In favour of 'preserved' is that this was the word eventually applied to the revival of the Talyllyn Railway in 1951, which is taken to be the first of the railway movements we're concerned with. But the man behind the revival of the Talyllyn, Tom Rolt, had originally spoken of 'perpetuating' the railway so it could continue to be a local amenity. A merely 'perpetuated' railway might not have become the major tourist attraction the Talyllyn is today.

'Preserved' implies something stuffed and mounted, removed from the real world, which brings us to another distinction: between preservation and conservation. Later in this book we will be meeting Dave Carr, Honorary Secretary of the Vintage Carriage Trust, which runs a museum on the Keighley & Worth Valley Railway. He uses 'preservation' as the wider term, meaning a holistic approach to the restoration of a vehicle, whereas conservation means an intervention on a particular part of it. The 'conservation' is likely to be more mindful of historical detail than the 'preservation'. 'But in the end', said Dave, 'you go for preservation because you want bums on seats.'

The bums on those seats often belong to tourists: all the preserved lines are dependent on tourism, which has grown, as leisure time has increased, since the founding of the Talyllyn, when the five-day week was not yet universal. Today, the Talyllyn is one of the Great Little Trains of Wales, a joint marketing scheme whose website is headlined, 'Experience the beautiful Welsh countryside: on little railways that are passionate about providing a warm welcome and an amazing experience,' and the technical railway element continues to be underplayed even as you go deeper into the website. Tom Rolt often complained about the tourists who visited the Talyllyn, but then he was not, as we will be seeing, a people person.

My own visits to preserved lines have prompted me to distinguish between those volunteers who are interested in *things* – engines, carriages, track – and those interested in people. I have often asked, 'Is this carriage BR Mark 1 or Mark 2?' I used to ask this because I didn't know; later on, I was curious to see whether the nearest volunteer to hand knew. Most didn't, but then I would often be referred to some guru of the line: 'Tom at the next stop will know. You'll find him in the carriage shed.' Or he would be at some other location not customer-facing, and Tom would know the answer to my question, and he would tell me ... not rudely, but pretty perfunctorily.

'Usual rules of thumb tally railway nuts at no more than 5 to 10 per cent of total visitor numbers,' writes Ian Carter in his chapter on preserved lines in *British Railway Enthusiasm*. The 'visiting enthusiasts tend to be despised,' he suggests, 'by the railway commercial managers' for their complaints that the railway is not 'authentic'. Carter lists some of these:

> that this locomotive never carried its current livery in service (lots of places), that in British Railways service this locomotive never carried the name with which its current proud private owner adorns it (lots of places), that to stick funny plastic faces on locomotive smoke-boxes for Thomas the Tank Engine weekends (almost everywhere) or to paint green and black engines red for Harry Potter weekends (lots of places) should see railway managers hanged, drawn and quartered.

Against the word 'heritage' is its connotation of schmaltz. And it has become a discredited word to left-wing ears. In his 1985 book *Living in An Old Country*, Patrick Wright depicts the rising heritage industry as a function of Thatcherite nationalism, but railways are barely mentioned. Then again, 'preserved' sounds like jam, but it better reflects the reality of these lines by suggesting something kept alive against the tide of history. We might say that the common denominator of the lines considered 'preserved' is that they are kept alive by rail enthusiasm, and in most cases that is the enthusiasm of the people who run them; but a small number are run on a purely commercial basis, in which cases it is the rail enthusiasm of the visitors that is important.

Since I seem to have come down in favour of 'preserved' over 'heritage', readers might be wondering why the latter is used in the title of this book. It's because some people consulted seemed not to know what a preserved railway was, whereas everybody could at least make a rough guess at the meaning of 'heritage railway'.

There is then the question of what to call the un-preserved railway. In the introduction I spoke of 'real-world' lines. We could speak of the 'national network', or the lines maintained by Network Rail, but that might become confusing, since eighteen of the preserved lines have a connection to the national network. Also, Network Rail has only existed since 2002, and might implode at any minute. I have already given 'the big railway' a run-out, but as a term it seems a bit infantile, which reminds me that the engines in the Thomas the Tank Engine books speak with awe (and distaste) of the lines beyond the Island of Sodor as the 'other railway'.

I think most of the time 'national network' would be best, because even though some of our preserved railways are groping towards each other, there is nothing like a national network of them.

The Parallel Lines

The scale of our preserved railway phenomenon is unique in the world. Yes, there are preserved lines in Europe – morbidly known as 'museum lines' – but nothing like as many. The question is why.

In public discourse about railways in Britain, there are two strains: veneration of bygone railways and condemnation of the present ones. It's hard to find praise of any particular modern service. In recent years there has been a big increase in travelling by train, but many of those new journeys are made by commuters in circumstances of overcrowded misery. So we have the 'parallel lines' identified by Ian Marchant in his book of that name, which is subtitled 'Or, Journeys on the Railway of Dreams'. He juxtaposes 'the real railway' with 'the railway of our dreams' (the first, he adds, is 'just shit'). The preserved lines represent the railway of our dreams, and their appeal is inversely related to that of the real railway.

A lot of the things I dislike about the real railway were encapsulated in the *Railway Magazine* of March 2018, in an interview with Sam Jessup, who 'has made a name for himself over the past decade as a leading light of railway design'. At the time, Jessup had just stopped working for Virgin Trains, for whom he had been a designer reflecting the 'values of the Virgin Group – welcoming, fun and different'. Asked what he was particularly proud of, he mentioned 'the advertising wrap for the *X-Men* train which, through the association with Hollywood superheroes, portrayed the message of speed and power'. Jessup also cited the 'humorous' (his word) message displayed in Virgin Trains toilets: 'Please do not flush nappies, sanitary towels, paper towels, gum, old phones, unpaid bills, junk mail, your ex's sweater, hopes, dreams or goldfish down this toilet.' Both of these were on the West Coast service. On the East Coast, he was most proud of the naming and branding of the new 'Azuma' trains, resulting in publicity heralding the new trains as follows: 'AZUMA. IT'S FAST. IT'S CRAZY FAST ... It's like a train and a laser made a baby while on holiday in Japan ...'

Sam Jessup is obviously respected in his industry,

but the 'humorous' approach of Virgin Trains has been depressing me for years. I think our railways ought to be dignified, presenting a fairly neutral backdrop, in which as many as possible of their passengers might feel at home. But when Virgin were involved in the East Coast route they set up a sort of red Portakabin on the main concourse of York station, from which they supplied passenger information while playing pop music from a ghetto blaster on its roof. It was as though they wanted to provoke me into calling them crass, just so they could call me elitist.

But I am sure that in thirty years' time, the amusing notices of Virgin will be widely romanticised, possibly even by me. Amongst railway phenomena widely hated at the time of their introduction but now widely loved we might include railways themselves, the coming of the 'Big Four' railway companies, British Railways and diesels.

The preserved lines also tend to be in attractive countryside, and their appeal is connected to nostalgia for a more rural Britain – a time when the countryside was the backdrop to everyday life, not just a Sunday jaunt. 'In most areas for at least two full generations all important comings and goings were by train,' writes David St John Thomas in *The Country Railway*. 'The pair of rails disappearing over the horizon stood for progress, disaster, the major changes in life: the route to Covent Garden and Ypres ...' We are talking here of emotion arising from a sense of place, and railway nostalgia really got going when the places were taken away, by the closing of a line.

In fact, railways were being closed from early on. 'The earliest known closure', writes Brian Hollinghurst in *The Pleasures of Railways*, 'is that of the Newmarket &

Chesterford Railway, between Chesterford and Six Mile Bottom, completed in 1846 and abandoned in 1851, when the Eastern Counties Railway decided to make passengers for the racehorse capital of the world travel via, and change at, Cambridge.' Between 1923 and 1939, 1,300 miles of passenger lines were closed. By 1960 another 3,000 had gone.

On 29 September 1935, the 1 foot-11½-inch-gauge Lynton & Barnstaple Railway – a 19-mile-long branch opened in 1898 from the LSWR station at Barnstaple to Lynton on the Exmoor Coast – was closed by the Southern Railway, its proprietor at the time. The line was one of the first railway victims of the motor car or, since this was tourist country, the charabanc. A thousand people turned out in pouring rain to watch the last train. The next morning, a Captain Woolf, RN (Retd), of Woody Bay, laid a wreath on the stop block at Barnstaple Town Station, with the inscription, 'To Lynton & Barnstaple Railway, with regret and sorrow from a constant user and admirer. Perchance it is not dead, but sleepeth.' If you change 'it' to 'she', this is a quotation from a poem by William Henry Furness, and fittingly literary, since the Lynton & Barnstaple – picturesque enough to justify a couple of observation saloons among the coaching stock – was a magnet for writers.

The Lynton & Barnstaple was written about so often that there is a Lynton & Barnstaple railway anthology, edited by David Hudson. This includes the 'obituary' for the line, which appeared in the *Railway Magazine* of November 1935. Of the line's opening, the obituarist writes, 'Festivities at Barnstaple ... began on 10 May 1898, when a Mayoral at-home in the Music Hall was attended by 400 guests. The formal opening took place on the following morning. The district seems to have made a public holiday of the opening, triumphal arches were erected ...' And this of the final journey ('a half excursion' from Barnstaple):

> In the evening, in the mist and the gloom, all Lynton seemed to have turned out to bid farewell to its railway ... the engines drew their train of nine coaches out of Lynton station for the last time to the accompaniment of their own shrill whistles, the playing by the [town] band of 'Auld Lang Syne', and the explosion of detonators.

The Lynton & Barnstaple was hymned by, amongst other notables, Henry Williamson and E. H. Shepherd. It also appears in *The Journal of a Disappointed Man* (1919) by W. N. P. Barbellion, pseudonym of a doomed amateur naturalist called Bruce Frederick Cummings (1889–1919), whose life was transformed in 1915 when he accidentally discovered that he had multiple sclerosis. (His doctor had given him a sealed letter to hand to an army medical board; after being turned down as unfit, he read the letter.) In the *Journal*, he records his decline and his discoveries in the natural world. He was born in Barnstaple, and here is a diary entry for 17 August 1911:

> Caught the afternoon train to C–– [this is Chelfham, prettiest station on the L&B], but unfortunately forgot to take with me either watch or tubes (for insects). So I applied to the station-master, a youth of about eighteen, who is also signalman, porter, ticket-collector, and indeed very factotal – even to

> the extent of providing me with empty match boxes. I agreed with him to be called by three halloos from the viaduct before the evening train came in. Then I went up to the leat, set up my muslin net in it for insects floating down, and then went across to the stream and bathed. Afterwards, went back and boxed the insects caught, and returned to the little station with its creepers on the walls and over the roof, all as delightfully quiet as ever, and the station youth as delightfully silly. Then the little train came around the bend of the line – green puffing engine and red coaches like a crawling caterpillar of gay colours.

Towards the end of his life, Barbellion wrote, 'Death can do no more than kill you.' It didn't even do that to the Lynton & Barnstaple Railway. In 1979, a Lynton & Barnstaple Railway Association was formed to revive the line. In 2004, a short section towards the Lynton end was re-opened, and there are plans to extend along the nine miles from the re-opened section to Lynton.

Railway Preservation Before Beeching

We have been speaking about the importance of a sense of place. Before the wholesale closures of the mid-twentieth century, however, railway enthusiasm had more to do with the sheer charisma of a steam engine, to which millions of words have been devoted, but the situation was pithily expressed by Rudyard Kipling in his short story '.007': 'A locomotive is, next to a marine engine, the most sensitive thing man ever made.' (The observation would have been even pithier if he'd left out the marine engine.)

Locomotives have always been preserved, but at first this was done haphazardly, and it seems reasonable to view those early preservations as milestones on the road to a glorious future: the preserved engines were not venerated simply because they were of the past, but because they were *important*.

The locomotive *Invicta*, built by Robert Stephenson in 1829, is said by some historians to be the first preserved locomotive. It worked on the Canterbury & Whitstable line (the 'Crab & Winkle'), which is regarded – usually by the same historians – as the first true passenger railway powered by steam. *Invicta* was put into storage by the South Eastern Railway in the 1840s and made celebrity appearances at many railway events.

Locomotion Number 1 – the engine that hauled the first train on the world's first public railway powered by steam, the coal-carrying Stockton & Darlington Railway of 1825 – was displayed outside the Hopetown Works of the North Eastern Railway from 1857. It had done some time as a pumping engine, and many early locomotives that have become famous endured a similar period of penance; the engine was then displayed at Darlington Station from 1892 to 1975. (Incidentally, just as the First World War was not called that at the time, *Locomotion Number 1* was originally called *Active*.)

In 1875 the North Eastern Railway held a pageant of historic engines to mark fifty years since the opening of the Stockton & Darlington Railway. This would be repeated in 1925, by which time the NER had become the LNER, and this time *Locomotion Number 1* appeared, albeit propelled by a petrol engine concealed in its tender, and with oily rags burning in the chimney to create smoke, if not steam. In 1975, when BR was in control, a 150th-anniversary pageant was held, and a British Transport Film was made of this event, with commentary from an urbane chap providing details of the contestants' achievements and vital statistics as though this were a beauty parade: '*Princess Elizabeth* 6201', he suavely intones, 'was also built to Stanier's designs. In 1936 she made railway history by doing the round trip from London to Glasgow and back at a speed of almost 70 mph over 800 miles ... Here is *Sir Nigel Gresley* streamlined Pacific class A4 number 4498, built with a taper boiler and three cylinders.' A replica of *Locomotion Number 1* led the parade. For the next fifty-year anniversary we are promised a 'global event in 2025 to mark the bicentenary of the Stockton & Darlington'.

Robert Stephenson's *Rocket* – winner of the Rainhill Trials held in 1829 to decide which locomotive should work on the Liverpool & Manchester Railway, (which would open four months after the Canterbury & Whitstable) – was donated in 1862 to the Patent Office's museum, which in 1885 became the Science Museum. Many replicas of *Rocket* have been made. A photograph in the archives of the National Railway Museum shows a static replica arriving at the Science Museum in 1935, on a lorry marked 'London & North Eastern Railway'. Schoolboys dressed like William Brown in caps and belted macs are crawling all over it, and there is a striking sense of a scene in the past twice over: relatively recent and further back, like the pluperfect tense. Today, *Rocket* and two of the replicas are in the National Railway Museum.

Sans Pareil (*Rocket's* rival at Rainhill) went to the Patent Office/Science Museum in 1864 after a spell as a pumping engine. The Science Museum had already got William Hedley's engine, *Puffing Billy*. This not to be confused with *Killingworth Billy*, which was recently discovered – after an 'archaeological' investigation of its components – to be, according to the *Railway Magazine*, 'the world's oldest surviving standard-gauge loco', having been built by George Stephenson for the Killingworth Colliery in 1816, not 1826 as had previously been thought. From the late 1880s, *Killingworth Billy* was displayed at Newcastle Station. Today, it is at the Stephenson Railway Museum at Middle Engine Lane, North Shields. (That a coal mine should have spawned the engine underlines the symbiotic relationship between steam locomotives and coal; it's as if cars had been built to transport petrol.)

Even by the end of the nineteenth century, rail enthusiasts, or 'railwayacs' (railway maniacs), were emerging as a distinct breed. They were often vicars, for reasons much speculated upon. (Is a railway timetable somewhat liturgical? Is the steam in a station akin to incense in a church? A more prosaic explanation is that vicars had plenty of free time.) What is taken to be the first rail enthusiast's society, the Railway Club, was founded in 1899 and wound up in 2008. I had joined a couple of years beforehand, and I had a feeling the club wasn't long for this world, because its annual dinners began with 'All rise for a toast to Her Majesty the Queen.' In August 1899, the Railway Club defined its objects as:

1. The provision of a bond of fellowship for the numerous 'Railwayites' in the country.
2. The study of the locomotive.
3. To gain information as to the stations and movements of locomotives.

4. To facilitate the meeting of those interested in locomotives.

The railwayacs' heroes were the chief mechanical engineers of the railway companies, whose glamour lay in their modernity, which for a while militated against preservation. In 1906 George Jackson Churchward, CME of the Great Western, scrapped the last two surviving broad-gauge engines built by the company, one of which was *North Star*, the engine that had hauled the first ever GWR train out of Paddington in 1838. Being broad, they took up too much space at the Swindon works. (In 1925 the Great Western tried to make amends by building a replica of *North Star*.) One broad-gauge engine survives, at the South Devon Railway Association museum at Buckfastleigh on the South Devon Railway, and while it is broad it is not very long, hence its name: *Tiny*. It's a four-coupled shunting engine with an upright boiler, and it was built for the *old* South Devon Railway.

The railways had been nationalised during the First World War. In an attempt to preserve some of the economies of scale this afforded, the Railway Grouping of 1923 shepherded 123 private railway companies into 'the Big Four'. The Big Four became preoccupied with stopping the drain of freight traffic to the rapidly expanding road network.

In trying to win over the public to the cause of the railways, the Big Four became adept at public relations. In fact, they invented it, at least as far as Britain was concerned. In 1925, a former Guards officer and journalist called John Elliot was recruited by Sir Herbert Walker,

General Manager of the Southern Railway, to improve the Southern's image: the disruption caused by the programme of electrification was annoying the bristling businessmen who were its main customers. Walker asked Elliot what he thought his job title should be.

'Well, sir', said Elliot, 'when I was on the *New York Times* in 1921, there was a man called Ivy Lee, who called himself "public relations consultant" to the Underground system in New York. So why not call me Public Relations Assistant?'

Walker agreed, 'Because no-one will understand what it means and none of my railway officers will be upset'. (Elliot later said that he never knew whether, having brought PR to the UK, he should have been commemorated by a statue in Parliament Square or 'publicly hanged on Waterloo Bridge'.)

In their publicity, the Big Four would be Janus-faced. On the one hand, they would present the railway as ultra-modern – symbolised by the development of bigger and faster locomotives, designed to haul ever more luxurious, therefore heavier, passenger carriages. Step forward Sir Nigel Gresley, CME of the London & North Eastern Railway, whose 'big engine' policy involved the refinement of the Pacific engine type. Gresley's Pacifics included *Flying Scotsman*, and culminated in the streamlined A4 class, of which *Mallard* set an as yet unbeaten world record for steam by running at 126 mph on 3 July 1938. There may have been only technicians, and no journalists, in the superbly named dynamometer car that measured the speed, but the feat was above all a public relations exercise, expertly managed by the LNER's urbane PR man, Cecil Dandridge. *Mallard's* run was partly about beating the LNER's rival on the Anglo-Scottish routes, the LMS, which had its own streamlined engines, the Coronation Class, designed by William Stanier, one of which, *Duchess of Hamilton*, stands a few feet away from *Mallard* in the NRM. (Both are, for the moment, sleeping beauties – or think of them as slumbering Arthurian Knights of the Turntable.)

But there was also a nostalgic aspect to the publicity of the Big Four.

Flying Scotsman was named after an Anglo-Scottish train first introduced in 1862. At the British Empire Exhibition of 1924, it was exhibited alongside a Great Western Railway Pacific named *Caerphilly Castle*, one of the sleek, long-boilered (therefore powerful) Castle class introduced in 1923 by Churchward's successor, C. B. Collett, and named after castles in the West Country. They would be followed by the King class, starting (after his permission had been humbly sought) with the current incumbent, George V. Then came the Hall class, commemorating venerable country houses (like Pitchford Hall, a Tudor pile in Shropshire,*) the GWR having run out of castles. These engines might as well have been called the English Heritage classes, and the idea was to make the GWR seem both patrician and patriotic, in contrast to the flat-capped Jack-the-Lads tearing about in army-surplus lorries stealing railway business. On the Southern, John Elliot would name engines after Arthurian knights (the King Arthur class), public schools (the Schools class), shipping lines (the Merchant Navy class),

*The engine of this name is today based at the Epping–Ongar Railway.

picturesque West Country locations covered by the Southern Railway (the West Country class), and heroic wartime RAF stations (the Battle of Britain class). The Merchant Navy class onwards were designed by Oliver Bulleid, who had been Gresley's assistant on the LNER, and was the last designer of flamboyant steam engines.* Steam traction was dying, and it was increasingly easy to imagine steam engines in museums.

The first railway museum at York – forerunner of today's NRM – was opened in 1928. It had a northern bias. Many of the exhibits had been used in the Stockton & Darlington centenary celebration, but they also included the GWR engine *City of Truro,* which had been unofficially recorded in 1904 as being the first engine to clock 100 mph. The man with the stopwatch had been the railway journalist Charles Rous-Marten, who campaigned for a national railway museum. The engines were displayed on plinths in an old loco-erecting shop. There was also, for a very rainy day, a 'small exhibits section', which displayed things like engineering drawings and signal block instruments.

It took nationalisation in 1948 to bring about a systematic curatorial programme. The British Transport Commission, which oversaw British Railways as well as other nationalised forms of transport, established a department with one of those honest old-fashioned names I'm nostalgic for: the Office of the Curator of Historical Relics.

In 1958, the Commission finally found a home for its relics, in an ex-London Transport tram depot in Clapham in south London. The Museum of British Transport was

*He designed his last engine for the Southern in 1949.

opened in 1961. A pamphlet about the place, written shortly after its opening, ascribes underdog status to the Museum: 'it was a little on the small side, the lack of rail connection meant that massive locomotives would have to be delicately manhandled into position and Clapham itself was no longer a fashionable area.'

In his report of 1963, Dr Beeching, with typical nihilism, recommended that BR quit the museum business, but that pioneer of preserved railways, L. T. C. Rolt, and the railway writer Jack Simmons led a campaign of opposition, and BR was persuaded to establish a national railway museum as a branch of the National Museum of Science and Industry. The new museum would be in an old locomotive roundhouse at York – a controversial decision, since all other national museums were in London. It opened in 1975, stocked with some of the Clapham collection.

Today, Clapham is if anything too fashionable for a museum, and there is a Sainsbury's on the site of the Museum of British Transport. The old York museum closed in 1973. I went there a few times as a boy. The locomotives were displayed on plinths, looking as dead as stuffed animals. You couldn't see how they'd got in or how they would ever get out and, back then, it didn't look as if they would ever have occasion to leave.

The major road hauliers had been nationalised along with the railways, but in 1951 road haulage was *de*-nationalised, and became once again a competitor to a railway network run down and exhausted by the war. Between 1948 and 1964 road vehicle mileage grew at 10 per cent a year. The Big Four had been experimenting with diesels, which were more efficient, and therefore cheaper, than steam locomotives, and in 1951 BR began a programme of dieselisation. In the same year, it also tried to create a more rational steam fleet by embarking on the building of its Standard classes. There would be 999 of these, and most lived less than half of a locomotive's expected lifespan of thirty years before being scrapped. About forty made it through to preservation.

In 1951, BR was making a small operating profit. By the mid-1950s, it had slipped into deficit, and freight traffic was being lost to road at a rapid rate. In 1955, therefore, came the BR Modernisation Plan – because this was now the technocratic era of 'plans'. For the railway romantic, the Plan was bittersweet. It proposed a big investment in railways and advocated only a few line closures, but it envisaged a push from steam to diesel and electric traction, which the romantics didn't like.

The Plan's subsequent failure to stem BR's losses meant that a flintier approach would now be taken. In 1962 Dr Beeching (see below) became BR Chairman. In that year – as he was donning his cowl while eyeing the underused branch lines – 2,000 steam engines were withdrawn, about one-tenth of the total fleet, and an unprecedented number for a single year. By the end of 1965, the Western Region of BR was steam engine-free.

In the case of the branch lines, the last running of a steam service might coincide with the end of the line, and so trainspotting took on a funereal aspect – and these funerals were well attended. Trainspotting ceased to be a matter of just seeing (or 'copping') an engine. Suddenly, time pressure was involved. It was a matter of copping it before it disappeared.

Steam engines are versatile performers. In the early twentieth century they could embody modernity; in the 1960s, they went in for bathos. Many were rusty and leaking steam from every orifice as they staggered through their final duties. If they were working alongside diesels their picturesque quality was thrown into relief, like a Victorian dandy standing next to a man in a boiler suit. There was a morbid fascination about this valedictory era that boosted trainspotting, albeit casting the spotter in the role of socially marginal grump. (In *British Railway Enthusiasm*, Ian Carter speculates that the derivation of the slang term for a trainspotter, a 'gricer', is a grouser.)

But some of the wealthier enthusiasts were not content to merely wave goodbye from the lineside. In 1959, Captain Bill Smith, an ex-naval officer, bought a tank engine: Class J52 68846. It had glamour by association, because it had been a pilot engine at King's Cross 'Top Shed', shifting the Gresley Pacifics and other thoroughbreds around when they were not in steam. In 1980, he donated the engine to the National Railway Museum, which out of gratitude made him an honorary life member. According to an article on the NRM website about 68846, 'This was the first mainline locomotive to be bought by an individual for preservation, and it made many people sit up and take notice – including Alan Pegler, who later bought 60103 *Flying Scotsman*.'

After a big payday in 1967, David Shepherd, the wildlife artist and painter of steam locomotives, bought two BR Standard Class engines, both of which are currently on the North Yorkshire Moors Railway. (Shepherd was a co-founder of the East Somerset Railway, whose signal box at Cranmore station contains a small exhibition of his work.) Also in 1967, a retirement home for steam locomotives was created in the form of Steamtown at Carnforth in Lancashire. This was based in an old LMS loco shed, and the original idea was to make the shed the headquarters of a steam network with the engines running over part of the old Furness Railway, in particular its Lakeside branch, which closed in 1967, and which had probed north to Lakeside Station on Windermere.

Steamtown was the brainchild of a GP, Dr Peter Beet, whose boyhood had been spent in the Lake District. According to his obituary in the *Guardian* in 2005, 'At night, he would listen to the sounds of locomotives on the West Coast. His first sighting of a Royal Scot engine in crimson lake livery had a lasting impact. From then on, he always wanted a big red engine.' He would acquire a big red engine, LMS Jubilee class 5690 *Leander*, in the early 1970s. It was one of many engines that found sanctuary in Steamtown. Dr Beet's original vision, of an operation involving shed plus line, was unfulfilled (too many roads in the way) but in 1973 part of the Lakeside Branch would re-open as the Lakeside & Haverthwaite Railway. In the same year, the opening sequence of the film *Swallows & Amazons* was filmed on the line.

Even without its own dedicated line, Steamtown became a tourist attraction and servicing point for rescued locomotives. In 1974, Sir William McAlpine became a shareholder in the venture, at which point he was the owner of *Flying Scotsman*, whose home became – for a while – Steamtown. But McAlpine wasn't the first rescuer of *Flying Scotsman*.

Alan Pegler was a flamboyant man who became

(briefly) rich for a prosaic reason: his family firm, Pegler's Valves, was sold to a firm called Northern Rubber. With his share of the proceeds he bought *Flying Scotsman* in January 1963 for £70,000. Pegler had been in love with the engine ever since, aged four, he had been lifted onto its footplate when it was displayed at the British Empire Exhibition of 1924. Coming home after the purchase, he sat down on the bed of his daughter, Penny, and said, 'I've bought a steam engine. I'm going to have her painted apple green [again] and we're going to have so many wonderful adventures in her.' There were also some misadventures. Ownership of *Flying Scotsman* would bankrupt Pegler, requiring him to move into a flat above a chip shop opposite St Pancras Station, and seek work as an actor impersonating Henry VIII at heritage-themed banquets. But Pegler is the reason we have all heard of *Flying Scotsman*.

I suspect, incidentally, that his remark about 'apple green' went over Penny's head: apple green was the livery of LNER engines, one of which *Flying Scotsman* had been, but at the time of Pegler's purchase it was in BR's green, which was darker, and sometimes called Brunswick Green, perhaps incorrectly, because Brunswick Green was really a GWR shade. Pegler, as already implied, was far from pedantic; nonetheless he was prioritising the repainting of the engine from one shade of green to another similar shade, and in this he was foreshadowing what would become a fraught and – to the lay observer – exasperating issue on preserved railways: what phase of an engine's life do you commemorate? When do you stop the clock?

Pegler had the support of the Prime Minister, the wily

Harold Wilson, currently positioning himself as anti-Beeching. Accordingly, Pegler had a special contract with BR which allowed him to run *Flying Scotsman*, hauling enthusiast's specials, on the national network until 1972. So the engine gained a lonely eminence, because scheduled steam running stopped in August 1968.

The last timetabled steam-hauled train was the 9.25 p.m. from Preston to Liverpool on Saturday, 3 August. Then, on 11 August, BR laid on the 'Fifteen Guinea Special', that being the price of the fare for a ride on a special valedictory steam train – involving four engines – running from Liverpool to Carlisle via Manchester and Rainhill. One of the four engines, a BR Class 5, 45110, ended up on the Severn Valley Railway. Thereafter, there would be no further steam specials (apart from those hauled by *Flying Scotsman*). BR was very firm about that, but it was in a Canute-like position, and in 1972 the ban on main-line steam was lifted. But for as long as it had existed, it had done the preserved railway business a good turn, in creating a demand for lines on which steam *could* run.

Beeching Versus Betjeman

In 1963 the British Railways Board published Dr Beeching's report, *The Reshaping of British Railways*, by which he meant cutting them by a third. Beeching was chairman of BR from 1961 to 1965, but the closures he recommended were implemented throughout the 1960s and early 1970s. Dr Beeching is to preserved railways what Guy Fawkes is to Bonfire Night – a villain who gives a pretext for a party – in that many of the closed preserved lines are re-openings of ones he closed, and the anniversaries of both closure and re-opening are usually celebrated. (But then the preserved lines will take any opportunity for a celebration. In late September 2018, the West Somerset Railway invited me to come and 'celebrate the end of summer'.)

One defence of Dr Beeching is the problematic one of 'only obeying orders', which directs attention to the person giving the orders. This was Ernest Marples, Minister of Transport from 1959 to 1964. Marples resembled a classic cad of the sort played in films by Terry-Thomas. He liked to be known as Ernie; he wore suede shoes, drove fast cars, used prostitutes, and briefed Beeching to implement railway retrenchment at the time when he (Marples) had an interest in companies constructing roads at public expense. If there was any doubt about his shiftiness it surely evaporated when, in 1975, he fled to Monaco to avoid prosecution for tax fraud.

His protegé, Dr Beeching, was a perfectly decent, even amiable individual. As we will be seeing, he appreciated the irony of being asked to re-open closed railway lines on behalf of preservation groups. It happened only twice, I think, but Beeching's grin might have slipped had he lived long enough to be invited to a hundred or so re-openings of lines he'd closed. There is a film clip of him making a tour of the national railway network, wearing a battered Tyrolean-style hat, which he raises in jocular greeting when some railway official makes an elaborate comedy genuflection to him as he walks through a ticket barrier. In *The Neologists*, a book about Britain's mid-life (or mid-century) crisis of the late 1950s and early 1960s, Christopher Booker describes Beeching as 'moustachioed but remarkably unsinister'.

Even though he could have played – on the strength of his balding and portly appearance – a stick-in-the-mud dad in some generation-clash film of the time, Beeching was associated with the cult of youth and dynamism that flourished in UK politics in the early 60s. Booker detects 'some link in terms of group psychology' between the teenagers' hostility to 'squares' and the rage of many journalists and politicians against 'middle-aged complacency' and 'stagnation'. Booker further relishes the irony of Harold Macmillan, who was lampooned by young satirists for his Edwardian manner, presiding over 'the greatest burst of radical activity since the years immediately after the war', citing the British application to join the EEC, the introduction of economic planning, university expansion and colonial liberation. Amid the economic setbacks and exchange rate crises that followed the 'never had it so good' years, Britain seemed to be falling behind its European rivals, and the fashionable question became 'What's Wrong with Britain?' Those who fretted away at the matter wanted solutions that were 'ruthless', 'drastic', 'dynamic'; they were impatient for 'competitiveness' and 'rationalisation'.

The *Sunday Times* Insight team, Booker notes, found Dr Beeching 'exciting': 'In a room filled with maps of the *doomed* railways, the doctor controls their *extinction* as *firmly* as he proposed it.' (The italics are Booker's.) Macmillan's successor, Harold Wilson, would take up the theme, in spite of being a pipe-smoking professional Yorkshireman (at least in public; in private he was a cigar man): 'This is our message for the Sixties – a Socialist-inspired scientific and technological revolution releasing energy on an enormous scale.'

In this climate, observes Booker, the word 'image' became important, and it was on Beeching's watch that plans were laid for the transformation of patrician British Railways into demotic British Rail, with a new, universal colour scheme of a depressing mid-blue, and the lion badge giving way to the modernistic double arrow logo. Beeching was not a railwayman when he was brought in to run the railways, and that was the whole point. His PhD was in physics, and he decimated the railways while on sabbatical from running ICI. In other words, he was 'scientific', which was one of Ernie Marples' favourite words. In 1959, opening the first stretch of the M1, Marples declared, 'It is in keeping with the bold scientific age in which we live.' Commending the Beeching Report to the Commons four years later, he said, 'it is not based on emotional thinking or wishes. It is based on actual research and scientific analysis.'

Beeching calculated that the least used 50 per cent of stations contributed only 2 per cent of revenue. He recommended that 2,363 stations (55 per cent of the network) and 6,000 route miles (30 per cent) be closed. Also, one-third of a million freight wagons would be scrapped, even though the Modernisation Plan had committed to the building of new marshalling yards to handle such wagons.

But Beeching sought in vain the 'profitable core' of the railway network. The 1962 deficit of £159m would only fall to £151m by 1968. He killed communities and threw 70,000 people onto the dole queue to save on annual expenditure a sum that in real terms was less than half the current annual railway subsidy. Since there was no concept of the usefulness of railways to the wider economy – of the so-called 'social railway' – (except among the romantics like Betjeman, who would not have put it like that), some very obvious depredations caused by Beeching were never even considered.

I wonder how many people died because Beeching turfed them off the trains and onto the roads, which he did at a particularly dangerous time, with plenty of drink-drivers at the wheel and seat belts optional. The peacetime peak of road deaths would come in 1966, when 7,985 people were killed – about twenty-two every day. On 2 May 1963, the Mayor of Chard in Somerset told the House of Lords: 'To annihilate the unprofitable, but extremely safe, railway system by increasing the lethal propensities of our ridiculous and costly road system will transfer the price to be paid in money to an account which will be paid for in blood.' (Chard had lost its passenger rail services in September 1962.) About half a million people died on Britain's roads in the twentieth century. As Gavin Stamp noted in *Britain's Lost Cities*, 'That is almost the same number as were killed in the Second World War (while the victims of railway accidents over the same period would merely fill a cemetery).'

Bus services were usually laid on to replace the trains, but people who had used the train did not necessarily want to avail themselves of this 'bustitution', or they couldn't, in the case of the infirm or women with prams. Most of the buses disappeared after a couple of years anyway. Beeching's aim was to whittle the railway down to the core of inter-city lines, and he believed people would drive to the stations on these lines – the 'railheads' – to board the trains. In practice they just drove all the way, resulting in congested towns and cities, so that whereas car use would be a reason to close railways in the early 1960s, it would be a reason to re-open them thirty years later.

Beeching was wrong, but he was absolutely convinced he was right, so he made no provision for the retention of the land over which the closed-down lines had run. BR was required to realise its value as soon as possible. This short-sightedness has thwarted many preservation schemes, but many have succeeded, and seldom has there been a more spontaneous and widespread refutation of any government policy. Many of the railway preservations would be instigated by students, even schoolboys, but then there was always something middle-aged about the youthful dynamism Beeching was supposed to represent.

I do not see why Dr Beeching should be regarded as any more perspicacious than the other pro-car people of the time, for example the author of a piece that appeared in 1960 in the journal of the Road Haulage Association, *The World's Carriers*:

> We should build more roads, and we should have fewer railways. This would merely be following the lesson of history, which shows a continued expansion of road transport and a corresponding contraction in the volume of business handled by the railways ... We must exchange the 'permanent way' of life for the 'motorway' of life ... road transport is the future, the railways are the past.

The wrongness of Beeching's thinking was surely proved by his second report, published in 1965. Its title, like that of its predecessor, was disingenuous: *The Development of the Major Trunk Routes*. Of the 7,500 remaining miles of trunk railway (or main line) 3,000 were 'chosen for future development'. The others were not discussed because they *had* no future. Under the plan, all train traffic to Wales would be funnelled through Swansea; large parts of East Anglia would have no railways; the East Coast Main Line would be closed north of Newcastle. The report was rejected as too drastic by the Labour Transport Secretary, Tom Fraser, and Beeching returned to ICI.

In December of that year, Barbara Castle became Labour's Transport Minister, and under her the concept of the 'social railway' was developed. It was enshrined in the Transport Act of 1968, which made available grants to subsidise unremunerative but useful lines. So the Beeching orthodoxy, with all its millenarian bombast, lasted no more than half a dozen years, although the actual closures continued, the last major one being that of the Alston to (the appropriately named) Haltwhistle line in 1976. Today, railways can only be opened; they cannot be closed.

This is because Dr Beeching lost the argument, and John Betjeman won it. Betjeman, poet, writer and broadcaster,* was Beeching's main adversary, even though they are not usually spoken of in the same breath, Beeching being seen as rigorously technocratic, Betjeman as shambling, eccentrically donnish. He was at the heart of the campaign against branch line closures. In 1954, when he was already the vice-president of the Railway Development Association, Betjeman was asked to become the president of the snappily titled Society for Reinvigoration of Unremunerative Branch Lines. He declined, becoming

* 'Poet and hack' was how he put it in *Who's Who*.

a vice-president instead, and did them the favour of advising they change their name to the Railway Invigoration Society, which they did. (It is today Railfuture, having merged with the Railway Development Association, and the eminent railway author Christian Wolmar is its Honorary President.)

Betjeman made several BBC TV films about railways. One of them (in 1962) was called *In View: Men of Steam.* It begins with him standing in the Swindon museum of the Great Western Railway, in front of the waterwheel-sized driving wheel of the broad-gauge *North Star* locomotive (well, the replica). Ignoring all the rules of presenting as usual, he speaks, crossly, and with folded arms, of 'cold-hearted Treasury nominees who will invent arguments against having railways at all – statistical ones – and of course, there are enemies of the railways in the road haulage industry.' In other words, the newly installed team of Marples and Beeching. The film is a rather rambling portrait of the Great Western. Betjeman goes along the Camerton branch of the Wessex main line (Bristol to Southampton), where it is as if he is attempting to wind up any Beechingites who might be watching: 'There's nothing like the peace of a branch line, where the porter has time to dream of the station competition.' The best-kept station competition, that is, and the porter is shown – at length – gardening.

Another of his railway programmes, dating from 1963, appears to have been called simply *Branch Line Railway*, and is a portrait of the line from Highbridge to Burnham-on-Sea, which had closed except for occasional summer specials in 1951. Even those trains had stopped in September of the previous year, so these were pre-Beeching closures, but Betjeman still makes Beeching the villain of the piece, here condescending to mention him by name. Over shots of Betjeman skipping about on the beach at Burnham (another provocation to the rationalisers) his voiceover intones, 'In ten years' time, when the roads are so full of traffic that we'll all be going by train again, you'll be grateful you still have a railway to your town. Don't let Doctor Beeching take it away from you.'

Betjeman's doom-mongering about cars was presumably regarded as a harmless aspect of his eccentricity. Certainly, the Shell oil company had allowed him to express those sentiments in his edition of the *Shell Guide to Devon*, but I suppose prophesies of traffic blight seemed far-fetched in 1934, when that book appeared.

In 1969, Betjeman wrote a poem called 'Dilton Marsh Halt', which manages to be a lament for a small country station on the Wessex main line that had *remained open*, Beeching's closure proposal being successfully resisted by the locals. The station had wooden platforms and no staff; it was a request stop. Passengers were directed by a sign to 'the seventh house up the hill', where a Mrs H. Roberts sold tickets on a commission basis. The poem asks whether it was worth keeping the halt open. The answer is yes, 'for in summer the anglers use it'.

Here is the last verse:

And when all the horrible roads are finally done for,
And there's no more petrol left to burn,
Here to the Halt from Salisbury to and from Bristol
Steam trains will return.

This was not one of Betjeman's most accurate predictions. The roads are *not* done for, and steam trains have never returned to Dilton Marsh Halt. (They don't need to, since it is served by modern diesels.) But the steam trains did return elsewhere. In *Britain's 100 Best Railway Stations*, Simon Jenkins describes Beeching as 'the enemy of memory'. But he was taking on some long memories. Britain invented railways and pioneered locomotive design; in 1849, Britain had more railways than the rest of Europe put together. Dr Beeching would come up against the enemies of the enemy of memory.

2

Some Pioneers

The Talyllyn Railway

The Booming of the Mountain Wind

The original Talyllyn Railway was a little over 7 miles long, and in the unusual 2-foot-3-inch gauge. It opened in 1865 to transport slate from a quarry at Abergynolwyn to the mid-Wales coast at Tywyn, from where the slate was transhipped to the standard-gauge railway that runs along the mid-Wales coast. (Talyllyn, incidentally, is also the name of the glacial lake at the foot of Cadair Idris 3 miles west of Tywyn, whose silvery sheen is beautiful but of an unnatural flatness, as when railway modellers use a sunken mirror to denote water.)

The Abergynolwyn quarry closed in 1946 but the owner of the railway (and the quarry), Sir Haydn Jones, was a rail enthusiast, and he promised to continue a passenger service on a couple of days a week for as long as he lived. The slate industry of North Wales had been in decline since the late-nineteenth century, and by the 1920s the line was mainly for tourists. A postcard of the 1930s shows sun-hatted trippers riding in an ex-slate wagon as if it were a funfair ride, and the message, written by a certain Pat to a certain Philip, reads: 'How you would love to go up the mountains in this funny little train. I wish you could see the engine, it would make you laugh.' A gravity ride was also available until the early 1930s. People would ride down the incline from Abergynolwyn to Tywyn in slate wagons. It was not quite as dangerous as it sounds, because the incline is at some points not noticeable; as the track plateaued, the wagons would have to be pushed to continue the ride.

The line was not included in the Transport Act of 1947, which nationalised the railways and other forms of transport. It is possible that officialdom had simply forgotten about it. In his account of preserving the Talyllyn, *Railway Adventure* (1953), L. T. C. Rolt marvelled at the way it had slipped through the net, when even such a ghostly thing as the Hereford & Gloucester Canal ('more difficult to trace to-day than Offa's Dyke') *was* included.

L. T. C., or Tom, Rolt (1910–74) was a sort of Betjeman with oily fingers: a trained engineer and prolific author. Rolt was descended on his father's side from minor Irish gentry and spent his early boyhood in a large modern villa built as part of a select development south of Hay-on-Wye in the Welsh Marches. The grounds of the house boasted a stable, paddock, coach house and tennis lawn. Tom's father, Lionel Rolt, was a swashbuckling type, keen on fast motorbikes. Various imperial adventures culminated – when Tom was aged ten – in a heavily loss-making investment, which required Lionel to find a house that could be run without staff. He settled the family in a stone cottage in Stanley Pontlarge, Gloucestershire. Unusually for a product of Cheltenham College, Tom Rolt served an apprenticeship with Kerr, Stuart, locomotive builders of Stoke-on-Trent, where his uncle was Chief Development Engineer. In *Railway Adventure*, Rolt listed his own interests (trains aside) as 'steam ploughing engines ... racing and veteran motor cars, beam engines, wind and water mills and boats of assorted shapes and size', and through his mechanical interests he expressed a philosophy.

When I'd been discussing preserved railways' patriotic affection for Britain's 'Finest Hour' with Bob Gwynne, the curator at the NRM, he'd made another, apparently contradictory, suggestion about the ideology of preservation: 'A lot of the people running them are basically syndicalists.' They don't look like syndicalists, being as a rule retired gentlefolk, still less *revolutionary* syndicalists, but what Bob meant was that they have (so to speak) taken back control of their own lives and localities in a not-for-profit way that accords with the tenets of libertarian socialism. Tom Rolt seems to embody this libertarian, socialist and localist approach, and his philosophy is expressed in his book *Narrow Boat*, which was written before the war and published in 1944.

The book, recounting a journey along the fading Midland canals, is as much polemic as travelogue, revealing Rolt as an angry young man. He is against mass production, the decline of craftsmanship, the promotion of quantity over quality. He dislikes seeing dartboards in canal-side pubs because they are killing quoits and bagatelle. In spite of his own mechanical bent, he wants to extract us from the 'mire of mechanised living', and *Cressy*, the barge in which he makes his tour, is an old wooden one. He likes canals because they go around rather than through. He is against fast travel, even by train, which to him signals a 'contempt for local environment, custom and tradition'. One reason he likes the bargemen is that they are often self-taught musicians. If they are illiterate that's no great loss: they have 'escaped the levelling influence of standardised urban thought and education'.

He is against not only mass-produced goods but mass-produced food ('foreign meat', 'cooking by can-opener'), although I'm not sure he was enlightened enough to do much cooking himself. This seems to have been left to his wife, Angela Orred, who accompanied him in *Cressy*. She had met Rolt in 1937 when he was running the very aesthetic Phoenix Green Garage at Hartley Witney, Hampshire. The breakdown vehicle was a modified Rolls-Royce Silver Ghost. Angela had swept into the garage in her Alfa Romeo, which needed a repair.

Rolt was a good writer. His feeling for language is expressed in his love of the place names cited as distance markers in a Bradshaw canal guide: Nip Square, Sheepwash Staunch, Maids Moreton Mill, Wainlode, Honeystreet. His slight tendency to piety is often relieved by humour. Approaching a lock at Cholmondeston on the Shropshire Union Canal, he wonders about pronunciation:

> I had felt tolerably certain it would be abbreviated to 'Chumston', until I was enlightened by an old canal lengthman at Nantwich, who not only pronounced it phonetically, but threw in yet another syllable, and stressed the last one but in such a manner that the ponderous word assumed the forcefulness of a rousing medieval oath: 'I was born Cholermondeston way.'

Oddly enough, industrial landscapes bring out some of his best writing. Here is Rolt on North Staffs pottery ovens:

> Shaped like gigantic bottles, blackened and squat, those that were belching dense coils of smoke from their necks looked as actively satanic as a volcano, but those that stood cold and dead had an appearance that was strangely ancient and oriental. They might well have been the pagodas of some temple to strange gods ...

Narrow Boat was read by a Robert Aickman, who is chiefly remembered today as a writer of ghost stories, but was also a conservationist, and he suggested to Rolt that they form a society devoted to the upkeep of Britain's crumbling canals. The two founded and began running the Inland Waterways Association in 1946, which saved, among others, the Stratford-on-Avon Canal. You'd have thought Rolt and Aickman would get on – both being conservationists who wrote ghost stories – but tensions arose over the scope of the IWA. Aickman wanted it to save every threated canal. For Rolt – a social miniaturist – this was too grandiose an ambition. The two fell out, each with factions on their side, and in 1950 Rolt's faction was ejected from the IWA.

By then he had the Talyllyn in his conservationist sights. In Spring 1949 he had taken the bus to the railway's Wharf Station at Tywyn from near the Talyllyn lake, where he was staying. It was a Wednesday – a running day theoretically – and early afternoon with the sky already darkening. He recalls the day in *Railway Adventure*. On the platform, Rolt was confronted with a sign, 'smudged by haste and rain: NO TRAINS TO-DAY'. He could hear a clanging sound:

> An engine was standing in the shed, and beneath her in the gloom of the pit between the rails, my eye caught the flickering light of a tallow candle. The sound of a heavy hammer striking unyielding metal was followed by a spate of rapid Welsh which, to my uninitiated ear, might equally have been instructive, argumentative or merely explosive and profane.

A few weeks later, Rolt returned with a locomotive engineer, David Curwen. Again, the trains weren't running, so Rolt and Curwen walked the track, all the way to the quarry at Bryn Eglwys. Here Rolt fantasised about the ghosts of old miners emerging from the dark, dripping tunnels. His first thought had been to resurrect the quarry, but as he listened to the 'booming of the mountain wind' amid the long cutting sheds he realised it was dead for good. The railway might be a different matter.

In his dislike of large-scale mechanisation, Rolt is in line with John Ruskin and his (Rolt's) near contemporary, George Orwell, and this strain of thought is not usually sympathetic to railways. John Ruskin hated all railways; Orwell damned the Metropolitan Railway for smothering Buckinghamshire with the Metroland estates. In 1949, Rolt hadn't yet started writing his railway books, and there had been a good deal of anti-railway sentiment in *Narrow Boat*. He had wanted to convert his barge to burn paraffin, but 'the railway company – characteristically – lost the case of vital parts' and he must wrangle on the phone with 'hopeless or apathetic clerks'.

If Rolt was resistant to the charm of the Big Four railway companies, he was not likely to approve of the new, nationalised concern. To him British Railways, with its national 'plans' and 'standard' engines, represented deadening uniformity. Here is our reliable old friend John Betjeman chiming in from the foreword to *Railway Adventure*: 'While those levellers, the clerks of the British Railways, are shutting down all the beautiful little country branch lines ... while they are killing the independent pride of the companies, while they are doing away with individual liveries and stamping their ugly emblem on every engine ...'

Rolt was not interested in endorsing a national network of anything. He was interested in that one line running through two valleys, located in – as Oliver Postgate used to intone at the start of *Ivor the Engine* – 'the top left-hand corner of Wales'.

Volunteer Platelayers Required

Haydn Jones died in July 1950. That summer, Rolt wrote a letter to the *Birmingham Post* proposing a rescue for the Talyllyn. Seventy people turned up at the subsequent public meeting, held at the Imperial Hotel, Birmingham. A committee was formed with Rolt as chairman and Patrick Whitehouse as secretary. (Whitehouse, who owned a building firm, would be involved in numerous preservations. He wrote fifty books on railways and co-hosted a BBC show, *Railway Roundabout*, which explained railways to children in unpatronising terms between 1957 and 1963.) Haydn Jones's executors arranged the transfer of the shares in the railway to a not-for-profit company created by Rolt, but he also created a Preservation Society, and he made sure the Society had control over the company. He wanted the people interested in the railway to have the upper hand over the people interested in the money. Most preserved railways have an operating company and a separate society of members. The Swindon & Crickdale Railway has 600 members; the Bluebell Railway has 10,500. You always have to be a member of a railway to volunteer, and if 10 per cent of the members are active volunteers, a line is doing well. Different power balances obtain, and there have often been tensions between the operating company and the members. In *British Railway Enthusiasm*, Ian Carter puts the Talyllyn set-up at the 'democratic end' of the preserved lines' constitutional spectrum. This spectrum 'moves through enthusiast-centred companies limited by guarantee, by shares or as plcs (the dominant forms amongst major lines) and sundry trust arrangements to the other extreme, where we find railways owned by private or public companies as more-or-less straightforward commercial investments.'

When Rolt's society took control of the line in 1951, it inherited two engines. As with most industrial railways, the engines were known to their keepers by numbers: there was Number 1, and Number 2, although they also had names: *Talyllyn* and *Dolgoch*. They'd always had their eccentricities. In 1866, a Captain Henry Tyler, making his initial inspection of the line for the Board of Trade, reported that Number 1 suffered from 'vertical motion', whereas Number 2 suffered from 'horizontal oscillation'. (He also discovered that the railway had the wrong loading gauge. In other words, the carriages were too wide for the bridges, so that the doors could not be fully opened on both sides when under the bridges. Tyler – who sounds like an easy-going man – agreed to an arrangement whereby the track would be slewed off centre as it went under a bridge, so that at least the doors on one side could be fully opened.)

When Rolt came in, only Number 2 worked. Soon after, he purchased a Number 3 and a Number 4 from the nearby and recently closed Corris Railway, which happened to have the same rare track gauge as the Talyllyn. This new pair were named *Sir Haydn* and *Edward Thomas*; a Number 6 when it arrived would be named *Douglas* (and if all this is beginning to sound a bit *Thomas the Tank Engine*, just wait). Readers may be wondering about Number 5. It was built by David Curwen using a Model T Ford engine, the transmission from Rolt's barge and an old slate wagon. It worked the winter passenger services (which ran on Fridays only) until 1953 when a failed gearbox caused it to be taken out of service and dismantled.

The office at Wharf Station Rolt found illuminated by 'a duplex paraffin lamp with a bowl of opaque glass ... from the chimney breast above the blackened cast-iron grate a moon-faced clock with a broken mainspring looked down silently.' A Victorian letter, written in beautiful copperplate by a ship's master and concerning a cargo of slate, had been half eaten by mice. Rolt's detailing of this decrepitude is such that you can't believe he didn't relish it, in a romantic way.

Rolt was entitled to occupy the office by virtue of being general manager of the line. He was in charge, in other words, which suited him. Like many advocates of communality, Rolt seems to have been anti-social and aloof. He certainly looks that way in photographs: tall and thin, like a Guards officer with reserved, hooded eyes. His saving grace was his second wife, Sonia (Angela having literally run away to the circus, to become a ringmaster). Everybody liked Sonia. Like her husband, she was artistic and technical; artisanal and upper-class. According to her obituary in the *Guardian*, she was 'born in New York to colonial parents'. After education in a convent school and the London Theatre School she began working in a factory wiring bombers. She then responded to a War Ministry advert for women to carry goods on the Grand Union Canal. They would be paid by the ton. She married an illiterate young boatman, George Smith, 'the Adonis of the cut', and the two carried coal throughout the war. The *Guardian* approved of the pair: 'he with blond curls, she with dark waves – tough, equal, modern'. George featured in an Ealing Studios documentary film called *Painted Boats*, directed by Charles Crichton, who would go on to direct *The Titfield Thunderbolt*, a film – about a group of villagers who save their local line from 'bustitution' – directly inspired by the Talyllyn just as many real-life preservation schemes would be. Rolt met Sonia at a screening of *Painted Boats* in Birmingham. She thought he was soft because his barge had a bath, but she joined his campaign to save canals and improve the conditions of canal workers. They married in 1952 and managed the Talyllyn together.

At the Talyllyn, Rolt's first idea had been to follow the example of the Ravenglass & Eskdale Railway in Cumberland and convert the line from narrow gauge to miniature at 15 inches. But that would make the line no better than the 'toy railways' found at seaside resorts. He did not want the Talyllyn to be 'a side-show for tourists'. Yes, it was tourism that had kept it alive, but the line's motto was *Ich Dien* (I Serve), and that meant – for Rolt – serving the local community. He came to dread the 'annual invasion from England' – worried that he might be an invader himself. He so disliked the 'loud, harsh accents of London, Lancashire, Birmingham or the Black Country' in the carriages, that he would travel with the locals in the guard's van. He didn't much care for the more genteel visitors either, especially a cut-glass woman who asked, '*Could* you tell me whether your little train is running?'

Rolt was pretty astringent about the volunteers, too. In early 1951, a sixteen-year-old schoolboy called Vic Mitchell read a small item in the *Railway Magazine* about the Talyllyn. It included an address for Rolt, to which prospective volunteers were encouraged to write. Mitchell told his friend, Alan French – who had a live steam model railway in his garden – and the two contacted Rolt. In June, as soon as the school holidays began, the pair packed their army-surplus rucksacks and travelled overnight from Paddington to Tywyn, where they found Rolt in his little office. Vic Mitchell describes Rolt as 'an authoritarian figure'. A *Daily Express* photographer was due to arrive and so Rolt recruited them for the shoot, which would demonstrate the scale of the volunteer effort Rolt was mustering – even though, in that second week of railway preservation, Vic and Alan were the only two volunteers he had. Rolt – in shirt and tie but with sleeves rolled up – posed with them as they pretended to be pushing a flat-bed wagon loaded with tools.

The coverage would bring many more volunteers to the line. Vic and Alan returned to the Talyllyn later in the year, but by now Alan had read another article about a preservation effort on another Welsh line – the Ffestiniog – which appealed to them as being longer, and better engineered than the Talyllyn. After another day's work on the Talyllyn, Vic and Alan told Rolt they were off to help save the other line instead. Rolt said, 'There is only room for one preserved railway in Britain.'

All of this was reported in the *Railway Magazine* of June 2018, in a profile of Vic Mitchell, who is now in his mid-eighties. He was steeped in railways from birth: 'Dad bought his first model railway before I was born, just in case I was a girl.' Even as a young boy, he was volunteering his services to the railways. He lived at Hampton in south-west London, and his parents' house overlooked the London & South Western Railway's Shepperton branch from Strawberry Hill: 'Cab rides were offered in return for sweeping platforms, polishing the brasswork in the urinals, cleaning out the goods shed ...' A Mr Waterman, the station-master at Hampton, was the sole guest at Vic's tenth birthday party. For much of his life, Vic was a dentist as well as a preservation volunteer (he stuck with the Ffestiniog). He is also the owner of the Middleton Press, whose books usually have their own shelves in the bookshops of the preserved railways. There are more than 600 titles, each chronicling in detail a stretch of line, so that typical titles would be *Uttoxeter to Macclesfield* or *Watford to Leighton Buzzard*. The aim is to cover the entirety of Britain.

Another early volunteer on the line, and one of the founder members of the Talyllyn Preservation Society, was Bill Faulkner. He became Managing Director of the Talyllyn Railway Company in 1964, by which time he had moved from Birmingham, where he ran a printing business, to Aberdovey, to be near the line. He had no railway background, his son Ian told me, but 'He was fascinated by anything mechanical.' Bill was a full-time volunteer driver on the line, and he always drove engine number 4, *Edward Thomas*, always while smoking a pipe. The full-time staffers on the railway might have developed a Pavlovian response to the two-tone engine whistle of Number 4, because it was Bill Faulkner who brought their wages in a battered leather bag.

Bill was the son of a shopkeeper, yet his fireman on Number 4 (therefore his subordinate) was sometimes Lord Northesk, with whom he shared an interest in cars: 'Lord Northesk,' Ian recalls, 'had an XK120 Jaguar, while Dad had an E-Type.' (Tom Rolt, founder of the preserved line, had also founded the Vintage Sports Car Club in 1934, and drove a two-seater Alvis 12/50 'Duck's Back'.) Bill Faulkner was a traditionalist: 'He wanted the line to stay as it was, so he opposed the extension to Nant Gwernol.' Ian describes himself as having been 'a railway child ... and my mother was a railway widow', but he remembers the railway with affection. 'I slipped away in the 1970s – moved to France – but I still get the *Talyllyn News*, and I follow it almost every day on social media.'

In 1951 the screenwriter T. E. B. Clarke (who would win an Oscar for *The Lavender Hill Mob*) visited the Talyllyn. He saw a sign reading 'Volunteer platelayers required', and this inspired him just as 'No Trains Today' had inspired Rolt. Clarke began writing what became *The Titfield Thunderbolt*, which was released in 1952 and has become the sort of origin myth of the preserved lines.

The film incorporates a version of an episode that had occurred on the Talyllyn. Coming into one of the stations, Brynglas, a driver – almost unbelievably called Dai Jones

– realised he'd miscalculated on his last refill of water, and his engine sputtered to a stop, so Jones, his fireman and the guard had to refill the tank with buckets of water collected from a nearby stream. The incident was also borrowed by the Reverend Wilbert Awdry, the Thomas the Tank Engine Man, for book number twenty-five in his series of twenty-eight titles: *Duke the Lost Engine*.

Awdry was of course a railway nut. In 1951, he'd wanted to go on holiday to Tywyn to look at the Talyllyn, but his wife (not interested in railways) had unfortunately booked a fortnight in Gorleston. But the next August, Awdry got in first and booked the family – wife, two daughters and eleven-year-old son, Christopher – into a guest house at Beach Road, Tywyn called *Monfa* (it's still there). They arrived on a Saturday evening, and a paradigm of sexism then ensued, as Brian Sibley writes in *The Thomas the Tank Engine Man*: 'Leaving Margaret to put the girls to bed, Wilbert and Christopher went straight to Wharf Station.' They were too late for trains, and there was no running on Sunday, so Awdry took the whole family back to the station first thing on Monday morning.

Mrs Awdry found the rolling motion of the little train made her seasick, so she took herself and the two girls off to the beach in the afternoon, while her husband took Christopher to the British Railways station where they 'invested in Cambrian Coast Runabout tickets'. At the end of the week even Christopher – who as an adult would take over the writing of the Thomas series – was beginning to feel that he'd had 'about enough' of trains of whatever kind, but his dad had fallen in love with the Talyllyn. He became a regular volunteer, working as a train guard. The Reverend Awdry *enjoyed* the motion of the train: 'First the engine would roll from one side to another, followed, like the joints of a caterpillar, by each coach all down the train so that it was often possible to see, at any one time, the various vehicles each leaning in an opposite direction to the one in front of it.'

When he first visited the Talyllyn, Awdry was half-a-dozen books into his Railway Series. The stories were always technically accurate, but the early books were simpler and more fairytale-like than the later ones, which were cluttered up with Awdry's concerns with railway politics and preservation. Like Rolt, he was anti-BR. In *Four Little Engines,* the tenth book in the series, he introduces two narrow-gauge tank engines: Skarloey and Rhenas, who are based on *Talyllyn* and *Dolgoch.* They operate on the Skarloey Railway, which Awdry would develop as the narrow-gauge line on his otherwise standard-gauge railway dreamland, the Island of Sodor. They had once served a quarry that has since closed down, but the quarry re-opens.

In May 1957, Awdry appeared – in his role as train guard – in a BBC documentary about the Talyllyn, presented by the two big guns of Welsh broadcasting, Wynford Vaughan-Thomas and Huw Wheldon, which seems like overkill. A short clip survives and is discoverable online intercut with some 'making of' footage. Vaughan Thomas and Wheldon are identically dressed in sports jackets and ties. They wield baton microphones and there is footage of the huge TV cameras which, with their high centres of gravity, look precarious mounted on a shaking flat-wagon as the train goes over the Dolgoch Viaduct. Tom Rolt appears in panama hat and cravat, smiling uneasily as Wheldon's attempts to interview a fireman (who was also Lord Northesk) are overwhelmed by the arrival of a flock of sheep. At the end of the clip, a disembodied voice – I think Rolt's – says, 'After the programme, our traffic virtually doubled overnight.' The film put the Talyllyn on the tourism map and enabled an investment programme that has continued ever since.

By the mid-1950s, the Rolts ceased to be full-time on the Talyllyn. They moved into Rolt's parents' house at Stanley Pontlarge, where their existence was more *Country Life* than *Country Living* – untainted by suburbia. One reviewer of *Landscape with Machines* remarked on the number of times Rolt used with approval the term 'remote'. The house, according to the *Guardian*, had 'canal-like conditions – paraffin lamps, open fires'. The Society for the Protection of Ancient Buildings advised them about its upkeep (literally, since it was falling down), and Sonia became involved with the organisation. She also worked with the History of Structural Engineers' history study group, and the Landmark Trust. Her husband, meanwhile, wrote many books. He died in 1974 aged 64; she in 2014, aged 95.

All Weathers and Conditions Bring Their Blessings

Tywyn is a small, sleepy town, predominantly grey, being made of slate, but not unattractive. It is not attempting to compete with the Snowdonian hills massing behind. I arrived on a Sunday evening; sea gulls were screaming in the sky. Like Awdry, I was too late for the Talyllyn trains, but I wandered down to the Wharf Station, which these days is more like a small complex, with bookshop and museum attached and, on the other side of the single track, a bungalow to accommodate visiting volunteers.

The track led away under a road bridge, then swerved out of sight. In the early days of the preservation, the volunteers at Wharf would watch anxiously as the engine disappeared and would be relieved when it returned. If it didn't, they went to look for it, and there would then come the job of putting it back onto the track with a length of rail levered on a sleeper. The track did look rather risibly narrow. I could have crossed it in a stride.

I walked into town and found the Salt Marsh Kitchen Bistro. It was not necessarily the only restaurant open, but the only one that *seemed* to be open. It advertised service from '5.30 p.m. til late', which in practice meant 8.15. As a tourist destination, Tywyn went through a rough period ten years ago, but the Salt Marsh, which is painted a sun-bleached shade of pale blue and boasts 'locally sourced' food, is one of a number of new, youthful – maybe even 'hipster' – venues that helped pull it through. The Talyllyn Railway has always been the tourist mainstay, however. 'If the Railway's running', said the Salt Marsh proprietor, 'then we're open,' and the Railway does run for most of the year. The proprietor was impressed by the volunteers: 'They come here for a holiday, but their holiday's working on the railway. They come from all over Britain, all over the world, and a variety of ages from twenties upwards, but some of them are very high-powered. I mean, you've got some of the blokes who held the nuclear codes volunteering on that line, and they leave it big bequests. Well, it's better than leaving your money to a cat home, isn't it?'

'What do they talk about?'

'They talk about the Railway – *all* the time.'

The next day, I returned to the Wharf Station, which was bustling in bright sunshine. Bunting, which I hadn't noticed the night before, fluttered along the platform. A blackboard propped outside the café advertised 'Narrow Gauge Breakfast' and 'Broad Gauge Breakfast'. An engine – Number 1 – was simmering away with a rake of red carriages, resembling a series of prettily wrapped Christmas presents. On the one hand the engine looked ungainly and misconceived, with its disproportionately big chimney and the bunker of coal in the wrong place (in front of the cab rather than behind); on the other hand it looked charming with its shiny, burnt-sienna paintwork. It was surrounded by photographers. The main difficulty of photographing an engine on a preserved railway is all the other photographers standing in the way.

I boarded the front carriage, an observation car that looked as plushly upholstered and historic as a Pullman saloon but was in fact knocked up quite recently. We pulled away under the bridge, the engine panting because of the ascent. We passed through Pendre, which means 'the edge of town', where the railway has its workshops, engine and carriage sheds. Now we were going faster and Number 1 was enjoying itself as we clove to the sides of hills in an incredibly green landscape. We passed under another bridge – very small – so our train was like a thread going through the eye of a needle. I could see the back end of Number 1 immediately ahead of us, and I could also see both driver and fireman because they were hanging out either side of the tiny footplate, as though they regarded the engine with distaste and would rather keep it at arm's length. 'It's just so hot up there on the footplate, so you hang out to get a bit of wind,' the fireman – a big bloke in a lumberjack shirt – later told me. 'I've probably lost half a stone this morning.' (The size of the cab means the fireman has to use a short shovel, so he is intimately involved with the fire.)

So many streams accompanied the line, or rushed underneath it, that I couldn't tell which was the one that had been raided by Dai Jones and his crew for the emergency re-fill, even when we were in the vicinity of Brynglas, where the event allegedly occurred. After Brynglas, the line is intermittently besieged by trees, and then comes the Dolgoch Viaduct with the raging white waters of the Dolgoch stream 50 feet below. There are three waterfalls in the gorge, and they are a reason to alight at Dolgoch Station. My little out-of-date guide book to the line was painfully conscious that rain is the typical weather of West Wales, hence: 'All weathers and conditions bring their blessings,' and hence also (and slightly ungrammatically), 'lucky is the traveller whose journey on the line, though marred by rain, can consequently see the Dolgoch falls in spate.'

I was sharing the observation carriage with two women in charge of three children. One of the children, a girl of about ten, said, 'I should have brought a book.'

'Why?' said her mother, as the girl had known she would.

'I'm bored,' said the girl, scowling at the heavenly vistas rolling by to the left (Snowdonia in bright sunshine). She was in a bad mood, and she knew she would annoy her mother by denigrating those vistas – so her remark was a kind of backhanded tribute to them. 'I could always go to sleep, I guess,' she said, and proceeded to pretend to do so, stretching out on the long velvet banquette.

We came to Abergynolwyn Station. The village – which lies three-quarters of a mile down the hill from the station – was built to house the quarrymen, or at least their families, because from Monday to Friday the men themselves slept in barracks at the quarry. The quarry lay beyond Abergynolwyn Station. A freight-only mineral tramway approached it, running to a spot called Nant Gwernol, but the actual quarry lay about a mile beyond Nant Gwernol, and 300 feet above it, at Bryn Eglwys. Cable-worked inclines connected Nant Gwernol to Bryn Eglwys.

Abergynolwyn's station building is long and low, like an attractive retirement bungalow in a sylvan setting. I asked the station-master whether the quarrymen would have thought of the line as beautiful. 'I'd have thought they would have hated it,' he said. 'Especially in winter'. (In *Railway Adventure*, Rolt noted that they had travelled in open-sided carriages, in which, even so, they were not allowed to smoke.)

Abergynolwyn was the original terminus of the preserved line, but in 1976, for the first time – and after much improvement of the line – passenger trains began running to Nant Gwernol. We were heading there now, the engine proceeding with a self-conscious, intrepid air along a densely wooded mountain shelf. The stroppy girl had now pretended to wake up from her pretended sleep, and she and her siblings had begun playing a word-association game with their mother. In light of the foregoing stuff about the Reverend Awdry the following is going to seem made-up, but it's not.

The mother said, 'Vegetable', and one of the children said, 'Supermarket.'

The mother said, 'Jurassic', and one of the children said, 'Heavy raptor.' (Lost on me, that one.)

Then the mother – after a glance through the window towards the engine – said 'Steam', and all three children chorused, 'Thomas!'

The Middleton Railway

The Middleton is only about a mile long, but any mile of railway in Leeds is going to be part of a historical tangle. It also doesn't look much, but this adds to its spirit of authenticity. The Middleton suits the kind of grey, overcast day on which I visited it, with rain swirling out of clouds of what might have been steam. I walked from Leeds Station through the austere red-brick streets of South Leeds, a place that carries a sense of absence. This was once a centre of locomotive manufacture, which in turn tells us that coal was nearby. In decades past, a railway traveller in some exotic place might be surprised to be reminded of Leeds by a brass builder's plate on an engine or carriage. For example, a passenger in silk pyjamas riding the *Train Bleu* in the 1930s might have seen a plaque reading 'Leeds Forge' on his or her luxury sleeping carriage, which would heighten their pleasure, since they were *not* in smoky Leeds, but speeding towards the blue skies of the Riviera. Usually, though, when a Leeds maker's mark was spotted – typically reading 'Hudswell Clarke', 'Hunslet Engine Co.', 'Manning Wardle', 'John Fowler' – the viewer would not be surprised. He'd probably be working in a goods siding or industrial railway, because the Leeds firms specialised in tank engines for that kind of work.

When I was growing up in York, I was intimidated by Leeds, with its heavy industrial pedigree. York's main industry – if you can call it that – was chocolate- making. Yes, railway carriages were manufactured in York, but engines hadn't been made there since Edwardian times. There was a geographical consolation, though. York is north of Leeds, so it was pure North Eastern Railway territory in the pre-grouping era, whereas Leeds was embroiled with the Midland, the Great Northern and (by virtue of being towards the west of Yorkshire) the London & North Western Railway, as well as the North Eastern. Until the late 1930s Leeds had three main-line railway stations, and it's one of those places requiring a special pull-out in any atlas of pre-grouping lines. Both the Midland and the Great Northern came to play a role in the Middleton story, but when the line started it had the railway world almost to itself.

In the 1750s Charles Brandling owned a coal deposit at Middleton outside Leeds. It cost him more to get the coal into Leeds than it did to excavate it. He proposed building a wagonway – rails along which vehicles could be pulled by horses – from Middleton heading north to staithes on the river Aire in central Leeds, a distance of about 4 miles. Those wagonways were common in the North-east, but Brandling's goes down in history because in 1758 he obtained ratification of his line by legislation, the first Act authorising a 'railway'. By making cheap coal widely available, this 3-mile link from the Middleton mines to the Aire brought industrial Leeds to life.

Gradually the Middleton became a railway without inverted commas, and for much of its life it was like a string frayed at both ends, with various connections at the northern end to staithes on the Aire, and various connections at the southern end to the coal deposits at Middleton. The wooden rails were replaced by iron-edged rails set 4-foot-1-inch apart and, in June 1812, the Middleton became the first to make commercial use of a locomotive. (The Stockton & Darlington, opened in 1825, was merely the first *public* steam railway). War has been described as the locomotive of history, and here that was literally the case, in that a shortage of hay and horses – causing people to think in terms of mechanical replacements – had been caused by the Napoleonic Wars. The engine was designed by a Matthew Murray. It was probably called *Salamanca* and was based on Trevithick's *Catch-Me-Who-Can*, an engine that ran in a circle – providing a 'steam circus' – in what is now Bloomsbury, London. *Salamanca* differed from *Catch-Me-Who-Can* in that it employed a rack-and-pinion system* to deal with the gradient between Leeds and Middleton. The success of Murray's Round Foundry in Water Lane fostered locomotive-building in South Leeds.

The first railway fatality is usually thought to be the death of William Huskisson MP, killed at the opening of the Liverpool–Manchester Railway in 1830, but in February 1813 a thirteen-year-old boy, John Bruce, was killed on the Middleton while running enthusiastically alongside the engine – so he was possibly also the first trainspotter. Another morbid first occurred in 1818 when the first locomotive boiler explosion killed the driver, who was blown 100 yards through the air into a field. Well, he was the man who happened to be driving on that day, but after his death the railway employed a regular, permanent driver – a James Hewitt, who therefore became the world's first

*Cog wheels fit into a toothed rack rail.

specialist train driver, until he too was killed in a boiler explosion in 1834. There now came one of the first of two periods when the Middleton, as though exhausted by its record-breaking feats, took a step backwards: the railway became horse-drawn again for a while.

Steam was reintroduced in 1866, with locomotives built by the Leeds firm of Manning Wardle. At this time, the gauge was also converted to the standard 4-foot-8½. The line began to be enmeshed into the national network, with junctions to the Midland and the Great Northern, and branches leading into a local scrapyard and an engineering works. By the early twentieth century, the Leeds end of the line no longer reached the river Aire. It had been cut back to a site adjacent to the Midland Railway goods yard, from where a loop also enabled it to serve a gas works.

To walk south across the Leeds Bridge over the Aire is to enter the industrial part of Leeds, or what *was* the industrial part. In 1888, Louis Aime Augustin le Prince poked a cinematic camera he had made through the windows of an ironmongers on the south side of this bridge (19 Bridge End – the window is still there) and shot the second-earliest cinematic film. It is like watching a slightly faulty clockwork toy: the carriages and pedestrians on the bridge move, then stop, then thrillingly move again. A glare of white sunlight slices in from the left. The main wagon in view is proceeding north and is heavily laden with what looks like bags of grain, and these had almost certainly been brought into South Leeds by train on the great fans of railway lines belonging to various companies. In particular there was the Goods Station of the

Midland Railway, whose site has now been obliterated by the sprawling Crown Point Shopping Village.

Today, the cityscape is full of lacunae: patches of scrub, crumbling red-brick buildings sprouting buddleia. The few new commercial buildings look plasticky and conditional. A few structures proudly survive from the days of industry, like the last men standing on the battlefield. One is the main building – a great mansion – of the Tetley Brewery, referred to on the map as 'The Brewery'. Today it's an arts centre, and the surrounding scrubland awaits redevelopment as part of the ambitious South Bank scheme – because Leeds is, on balance, a boom town these days, the wealth in IT, retail and financial services.

I walked around South Leeds armed with a street map dating from 1906. The Middleton Railway was on the map, snaking its way through the Dark Satanic Mills and labelled, in the pleasingly formal language of the map, 'Mineral Railway'. At this time it terminated, as mentioned, near a gas works. After getting lost many times, I located what I thought must have been the terminus. I couldn't be sure because it was in a street that – like many in South Leeds – had lost its name, or at least any indication of its name, in all the years of post-industrial turbulence. But I think I'd found the right spot because there is still a gasworks on one side of the street. When I saw this, the whole fumaceous world of the Middleton Railway came back to life for a second.

The Hard Sell

In 1947 the Middleton Estate and Colliery Company became part of the National Coal Board. With nationalisation came rationalisation, and the Middleton line began to be closed in stages.

In 1959 the situation was as follows. The northernmost part of the line, going into central Leeds, had been closed. The southernmost part of it was operated by BR, which used this stretch to carry coal away from the Broom Pit, using the junction with the ex-Great Northern line, which itself of course had now become part of BR. This left a short, middle section of the line, which was owned by the engineering works, Clayton, Son & Co., and there was a branch off this line to their works. There was also the branch to the above-mentioned scrapyard. This middle section was eyed with interest by the Leeds University Union Railway Club. At the time, many universities had a railway club (the only one remaining, as far as I know, is at Cambridge University), but the Leeds one seems to have been particularly dynamic. It wanted to take over and run a stretch of line as a kind of working museum. I haven't been able to find out the identities of the undergraduates involved, but I picture them as part of that gentle, retrospective strain of the late 1950s and early 1960s: duffle-coated, Aldermaston-marching, trad-jazz-fancying.

The club approached Clayton's and asked if it could use their line – which particularly appealed because of its distinguished history – for this purpose. The answer was yes, and so an offshoot of the club, the Middleton Railway Preservation Society (MRPS), was formed. A leading light of both was a lecturer at Leeds, Dr Fred Youell. He was a railway fanatic, as was his wife, Susan, who had been a maths student at Leeds. They had three children, Harriet, Sarah (whose middle name was Claudia, as a tribute to the chairman of the Great Eastern Railway, Claud Hamilton) and Matthew, who was named after Matthew Murray, builder of the first engine on the Middleton.

Some of the Leeds University people were into trams, which were being phased out at the time in Leeds. The MRPS acquired a few worn-out tramcars, and Hunslet Engineering loaned them a 1932 0–6–0 diesel, No. 1697, which description sells the engine a bit short, since it was the first diesel to run on a main line in Britain. On 18 June 1960, the MRPS used this first diesel to haul a coach.

This coach had just been decommissioned from what was possibly the world's first passenger railway full stop. ('First' fatigue may be setting in, so I will make this quick.) It was an enormous double-decker, resembling a tramcar, and it was from the Swansea & Mumbles Railway, which had been opened in 1806 to carry limestone from Mumbles in Swansea Bay to Swansea. In 1807 the line had begun carrying passengers, hence its claim to fame. At the time, it was horse-drawn and officially known as the Oystermouth Railway, which might have influenced the nomenclature of the Far Tottering and Oyster Creek novelty railway built by Rowland Emett for the Festival of Britain in 1951. In 1877, the Oystermouth line was converted to steam, and in 1928 it was electrified, using overhead transmission. The Mumbles Railway was both amusingly rickety and progressive, perpetually keen to get on to the next form of traction – if it were still around, it would probably be hydrogen-powered – but it closed in 1959, hence the appearance of its coach on the Middleton.

On Monday, 20 June (as reported in the next day's *Yorkshire Post*), 'Dr F. Youell, ... wearing academic dress, took over the controls of a light engine [the Hunslet diesel] and gently pulled away a train full of eager children.' The children were in the Mumbles coach ... and so the Middleton became the first standard-gauge railway to run a service. The passenger service only lasted that week – it was the Leeds University rag week, and rides were advertised 'at your own risk'.

A service would continue thereafter, but a *freight* service. The society persuaded Clayton's, the engineers, and Robinson & Birdsell's, the scrap merchants, to let them run goods trains carrying heavy steel and scrap between their two business premises and the transhipment sidings at the junction with the Midland. As payment for this honour, the Middleton Society agreed to insure the line, and provide trains whenever required. The freight service began on 20 September. Three empty wagons were taken to Birdsell's; two went out the same day – full of scrap to the sidings. On average, two trains a day were run, some comprising twenty wagons. A variety of old diesels of Leeds manufacture were used, and the odd steam engine. On weekends, the public were invited to watch, and non-student volunteers were recruited to fill in during university vacations. The MRPS had no stations at this point, and the timetable was out of kilter with what would become the preserved railway norm. Services might run in the evening, and August was the slackest rather than the busiest month, because August is always slack for freight.

Picture Freshers' Week at Leeds University in 1963 or so: there are stalls for Chess Club, Film Club, Christian Union etc, and then a stall for the Middleton Railway Society.

'What do you do?' asks a gauche young fresher,

thinking the activity might involve photographing trains or collecting numbers.

'We drive heavy industrial freight trains,' comes the reply. 'Do you want a leaflet?'

The Broom Pit closed in 1967, and the society – now called the Middleton Railway Trust – eventually acquired the southern end of the line. In 1969, with the demand for the freight service declining, the Middleton began catering to passengers again, and it instated its current termini, at Moor Road and Middleton Park.

The excellent guide book to the Middleton, from which much of the above is derived, was published in 2004, sufficiently long ago for the author to have boasted that an episode of Jimmy Savile's radio programme, *Savile's Travels*, was recorded on the line. The Middleton also provided what the booklet calls 'the sinister railway depot' in the BBC's apocalyptic, not to say hysterical, drama series of the mid-1980s, *Edge of Darkness*.

This is the cue for a lamentation in the line guide. The 'industrial scenery' of the Middleton makes it 'a hard sell when compared to other preserved lines with olde worlde stations and beautiful scenery.' Another aspect of the railway's grittiness: it is plagued by vandalism. (Whereas in 1812, the danger to *Salamanca* was from the more justifiable vandalism of the Luddites.) In the early 1970s, the creation of the M621 threatened the line, although Dr Youell thought a level crossing would take care of that. In the event, a tunnel was required to be built, because 'a railway used at least once a year cannot be compulsorily purchased.'

Dr Youell died in 1998; his ashes were buried under the track at Moor Road Station, and there is a plaque to his memory. Susan Youell had become Secretary to the Association of Railway Preservation Societies, which became the Heritage Railway Association. She died in 2017, and an obituary appeared in *Heritage Railway* magazine: 'She inspired hundreds through her teaching [and] ... enthusiasm for life and no-nonsense "can-do" attitude, especially for girls in a man's world.' In the last fortnight of her life she visited the Keighley & Worth Valley Railway.

The principal building of the Middleton's northern terminus, Moor Road Station, is plain red brick. The widespread use of red brick in Leeds denies the city the beauty of its neighbour, Bradford, which is mainly of gold-coloured millstone grit. The station forecourt was a car park, on which traffic cones and coal bunkers attested to new and old transport modes. On the edge of the car park, beneath an open-sided prefab shelter, stood an enormous tank engine, seemingly made entirely of rust. This was *Picton*, made by the Hunslet company for export to a Trinidadian sugar plantation. The main station building had the air of a cheerful community centre, being brightly lit with a counter for tea and coffee, model railway, books and leaflets on display. The gent selling tickets was very friendly in a northern way: 'Oh, don't bother about times, the train just shuttles up and down all day.'

To reach the platform you walk through the museum, with an impressive display of small industrial locomotives, mainly made in Leeds. The engine waiting at the platform was *Brookes Number 1*, made by Hunslet in 1941 for Brookes Chemicals. *Brookes Number 1* is a saddle tank (the water tank is draped over the boiler), but a few years ago it suffered for a while the indignity of being fitted with side tanks so that it could become 'a licensed "Thomas" character'. The junction with what was the Midland Railway was visible from the platform behind locked gates. It could still be used, allowing the Middleton's antique engines to escape into the real world (Leeds or Sheffield), but it *hasn't* been used since 1990.

There were two carriages – ex-parcels vans of the Southern Railway resembling barns on wheels. Both were full, mainly with cheerful and noisy families. We started away, passing some spindly trees with bin-bag tatters in them. A council estate was on one side, an industrial estate on the other. The guard, who worked in IT, was about a third of the age of the ticket seller, but just as friendly, and there was no resentment in his voice when he said (after I'd told him I was a writer), 'Yes, a lot of the media-type people don't understand the importance of this place.' We passed some five-a-side pitches belonging to the John Charles Centre for Sport. The guard explained, 'John Charles – Leeds United and Juventus. One of the first English players to go abroad. Gentleman Jim, he was known as.'

The termini are the only stops on the line, and we were now at the southern one: Park Halt, which is on the edge of Middleton Park, where the remains of coal excavation can still be seen. The trees of the park looked a bit sickly, but this was ancient woodland, as historically significant as the railway, which of course has a project: an extension into the park, as though yearning to make contact with the black gold again.

The Bluebell Railway

Some Geography

Most of the 11 miles of the Bluebell Railway are in West Sussex, but the principal and southernmost of its four stations – Sheffield Park – is in East Sussex. From where I live any part of Sussex is hard to get to by car, because London is in the way. But I did drive to the Bluebell, rather than utilising its perfectly good connection to the national railway network at East Grinstead. I did so because of the weather. Snow was falling, and Radio 4 had warned against all travel, but I thought my car was less likely to be immobilised than the Southern rail services to East Grinstead. My attitude to my car is that I disapprove of it, but I also like it, and modern trains do not seduce me away from it as readily as those on the preserved lines, whose old carriages have car-like virtues, namely wide, comfortable seats, big, openable windows and the possibility – in compartments – of silence and privacy.

I was penalised for my environmental irresponsibility, however. With snow coming down fast from a violet sky, I found myself lost, either on or in the vicinity of the B2028, by which I had hoped to approach Sheffield Park Station. As I parked alongside some snow-covered village green, wondering which turning to take, my trusty *Railway Atlas, Then and Now* (opened on the passenger seat) disclosed some more definite geography.

What became the Bluebell Railway started life as the Lewes & East Grinstead Railway, which ran broadly north–south, with East Grinstead the northernmost point and Lewes the southernmost, where it intersected with a line to Brighton. The line was sponsored by some local landowners, including the 3rd Earl of Sheffield,

Henry North Holroyd, who required it for agricultural purposes – and cricketing ones. The Earl, while not a very good player himself, was keen on the game (he established the Sheffield Shield, the principal domestic competition in Australia) and the only time the L&EG would be busy was when he held a cricket match at his mansion, Fletching. The Australian team tended to start their tours at Fletching, but the Earl also hosted village games. There is still a cricket pitch at Sheffield Park Gardens, and it is still played on by village teams. The Earl did not approve of the use of boundaries for club cricket, which I'm sure is something the *Test Match Special* team could talk about for hours if rain stopped play. The gardens are open to the public through the National Trust; the house is in private hands.

The authorising Acts for the railway, passed in 1877 and 1878, included the following clause: 'Four passenger trains each way daily to run on this line, with through connections at East Grinstead to London, and to stop at Sheffield Bridges, Newick and West Hoathly.' Remember that, because it will become significant. The line opened in 1882. It passed into the hands of the London, Brighton & South Coast Railway, then the Southern Railway, then BR.

In 1954, before Beeching loomed into view, BR proposed its closure, and this went ahead on 28 May of the following year. But then an Ealing Comedy-like scenario unfolded. Enter Miss Margaret (Madge) Bessemer. She had an affinity, perhaps, with things mechanical, being the granddaughter of Sir Henry Bessemer, that alchemist of early industry whose Bessemer Convertor turned pig iron into steel. Madge Bessemer – we might imagine her played by Margaret Rutherford – was a teacher at a village school, a parish councilor, a captain of Girl Guides and a commandant of the local Red Cross. She lived on the family estate near Newick Station to and from which she had secured from the railway her own private footpath. She liked to pick wildflowers on the embankments of the line, and it is thought that it was Miss Bessemer's idea to call the saved line the Bluebell.

But we are jumping ahead of ourselves. Indignant at the closure of the line, Miss Bessemer came across the above-mentioned clause, requiring those four trains a day in perpetuity, which meant that the closure was illegal until a revoking Act was passed. Pressurised by Miss Bessemer, who was abetted by her local MP, Mr Tufton Beamish (a name that seems to belong in *The Titfield Thunderbolt*), BR grudgingly reinstated a service, with trains stopping at the places along the line that had been listed in the clause. Bluebell people call this 'the Sulky Service', since its four trains were deliberately timed to be not much use, starting too late and ending too early in the day. This lasted until 16 March 1958, when the founding Act was repealed, and the line was closed again. But the last train on that day attracted a big crowd, and amongst it were people who were not so much sulky as determined.

Principal among them were three schoolboys, David Dallimore, Martin Eastland and Alan Sturt, all at Brighton, Hove & Sussex Grammar School, where the three had started a model railway society. They were all railway enthusiasts – trainspotters, indeed – but as David Dallimore (the only survivor of the three) recalls:

> On the whole, railways were not very interesting around Brighton, which was mainly served by electric multiple-unit trains. There was one exception though – the Lewes-to-East Grinstead line, where steam traction still held sway. A bizarre collection of locomotives was used on the line, ranging from diminutive Victorian tank engines right up to modern Pacifics. We spent many happy hours together cycling to the line and watching trains – we couldn't afford to travel on it.

They viewed the closure of the line as 'an act of vandalism' and, inspired by the example of the Talyllyn and the Ffestiniog, they believed it could be saved.

They already knew Margaret Bessemer and had been round to tea at her cottage. David Dallimore recalls 'a very cosy front room. She was well-educated, well-spoken with a very sharp mind. She was well into her eighties at that point.' Shortly after the closure of the line, David Dallimore had been holidaying in Wales on the Ffestiniog Railway, when he found himself sharing a carriage with a man who turned out to have a profound connection with the Lewes–East Grinstead. This was Bernard Holden, who'd been born in the station house at Barcombe, on a stretch of the line that has not been preserved. His father was the station-master, and his predecessors had worked for the LBSCR since 1840. Holden himself had joined the Southern Railway in 1925, working as a ballast train clerk. He rose quickly through the ranks. 'During the Blitz,' according to Holden's obituary in the *Daily Telegraph*, he had 'plotted the routes for trains round the bomb craters. At one point the only way to reach London from the south was to take the train to Balham, then the Tube; every other rail line was blocked.' Holden then had many military railway adventures in India. When David Dallimore met him, he was back working for BR. Holden was useful to boys because he was an adult – and given that he was working in the BR General Manager's office at Liverpool Street he ought to be taken seriously *by* BR.

The boys had been researching the chances of purchasing and operating the line. They decided to call a public meeting, to be held at the Church Lads' Brigade Hall in Haywards Heath on 15 March 1959. Martin Eastland booked the hall out of his own money and hoped enough people would turn up to recoup the cost – and indeed the train fare home, because he didn't have that after paying the final instalment on the hall. David Dallimore did not attend the meeting, acquiescing to the wishes of his father, who ran a local bus company and would rather his son were not publicly associated with what might become a rival organisation.

About a hundred people turned up. The meeting was chaired by Bernard Holden, the Bluebell Railway Preservation Society was created, and a committee was formed. It included Alan, Martin, Bernard Holden and Chris Campbell, a student of banking at Carshalton Tech College who'd got to know the Bluebell because he had an uncle and aunt living near it. Chris Campbell's particular interest was branch lines, especially those of Devon, which he'd visited on boyhood holidays. He'd first got interested in railways as a result of visiting signal boxes in North Wales and Cheshire with his grandfather, who sold life insurance, often to signalmen.

Originally, the Bluebell Society wanted to re-open the

whole line, and run it as a normal service with diesels, but it became apparent that there wasn't the local demand for that. This would be a common syndrome in the early days of preservation. Speaking of closed lines generally, Jonathan Brown wrote in *The Railway Preservation Revolution*: 'the local population got used to life without it and bought cars.' Preservationists might have been able to reverse a line closure in the early 1960s; they couldn't overturn the idea that railways were retrograde. Today, things are different. Here is Simon Jenkins, writing in the *Guardian* on 29 November 2017:

> Old railway lines have been steadily reopening. They have been revived everywhere from Manchester to Lincoln, Paisley to Edinburgh. Sections of London's Overground and Thameslink were formerly cut by Beeching ... Most sensational is the contribution of the heritage rail sector. If you want to see a crammed station try Loughborough on the Great Central, or Sheffield Park on the Bluebell, or Porthmadog on the Welsh Highland ... These are not hobby railways, as Grayling's Whitehall colleagues like to dismiss them. They were the guerrilla irregulars, who for half a century battled to reverse Beeching.

The Bluebell Railway Preservation Society settled for creating a volunteer-run steam railway operating between two of the middle stations on the old Bluebell: Sheffield Park and Horsted Keynes. The Society secured from BR Southern Region a lease on this middle stretch at the affordable rate of £2,500 per annum, and the line that commenced operations on 7 August 1960 ran over 4 miles from Sheffield Park to a spot called Bluebell Halt, a hundred yards short of Horsted Keynes. They couldn't quite go into Horsted Keynes because BR was still in residence there, operating another branch – this one electrified – south to Brighton via Hayward's Heath. Note the date: 7 August 1960. The Middleton Railway Preservation Society had run its first train on 18 June of that year, but the Middleton was using a diesel, albeit a very early one, whereas the Bluebell was using steam, so it can – and does – claim to have provided the first 'preserved standard-gauge, steam-operated passenger service'.

By 1960 the young founders had drifted away from Sussex, but they kept in touch with the Bluebell. Bernard Holden would become the 'father of the line', first as signalling engineer, then president. In 1992, when he received the MBE for services to railway preservation, the Queen urged him to 'keep up the good work.' He was 84 years old. He died in 2012 aged 104.

But to return to geography. In April 1962, the Bluebell opened another station on its 4-mile line. It was just south of Bluebell Halt, and it was called Holywell Halt. It was opened by Dr Beeching, who lived nearby at East Grinstead. Odd to think of him *opening* a line.* He'd been invited to open the halt by Chris Campbell, who was by now not merely a student of banking but an actual banker at a London bank where Beeching was a client. In 1962 Beeching was not yet the axeman, merely the chairman of British Railways, and Chris Campbell told me, 'He was

---

*In 1969 Beeching was 'guest of honour' at the reopening of the South Devon Railway, his presence, according to Richard Herring in *Yesterday's Railways*, a matter of 'unconcealed irony'.

a very straightforward person. Most people who knew him liked him.'

'Yes,' I said, 'but surely you, with your interest in branch lines, must have started to dislike him when he published his notorious Report?'

'Not really,' said Chris. 'He was a decent man. You know, he even cut his own local line: East Grinstead to Tunbridge Wells.' (By this I think Chris was suggesting that Beeching was incorruptible.)

The notorious Report of 1963 recommended the closure of the electrified branch to Horsted Keynes – and Horsted Keynes Station – leaving the Bluebell's line severed from the national network. In fact, it might be more accurate to say that Beeching removed Horsted Keynes from the national network rather than closing it, because the Bluebell had been camping in the station since late summer 1962. In 1963 it moved in properly, closing those earlier staging posts, Bluebell Halt and Holywell Halt. In sum, then: BR (pre-Beeching) closed a stretch of line; the Bluebell people re-opened it. Dr Beeching agreed to perform the opening ceremony at a brand-new station created on this formerly closed stretch of line. He then closed another station that had remained anomalously open on the closed line because it also served another line that remained open until Beeching closed it.

Ten years later, the Society then began inching back north towards East Grinstead, where it would be able to reconnect with the national network. It did this first by acquiring the freeholds of the sites of the two stations between Horsted Keynes and East Grinstead, namely West Hoathly (which had been demolished) and Kingscote, which survived as a structure – mock-Tudor, to attract genteel commuters – despite having gained fame for being the most obscure station on the LBSCR. The Society didn't rebuild West Hoathly but re-laid the track through its ruins, and by 1994 had reached Kingscote, after re-laying the track through the Sharpthorne Tunnel, which is the longest tunnel on any preserved line. In 2013, East Grinstead was reached, which is an easy thing to write, but represents a major feat of civil engineering, as we will be seeing.

The Sulky Service

Alongside the village green, a man was climbing into his car. Winding down my window – thereby letting the snow in – I asked him if he knew of Sheffield Park Station. 'I'm going that way,' he said. 'Follow me.'

Escaping the orbit of the village green, we entered a snowy woodland. It was humiliating to be guided in this way, especially since I'd driven to the Bluebell before, but now it seemed to have receded deeper into Sussex, deeper – perhaps – into the past.

(Its remoteness is the reason it closed, although sometimes remoteness can be a reason for a line to remain open. In June 2019, I read in an article in the *Railway Magazine* about the 13-mile Alston–Haltwhistle line in Cumbria that 'It was the remoteness of communities in the South Tyne Valley that led to the railway which served them surviving long enough to be become one of northern England's last rural branches.' It did close, though, in 1976, once a so-called 'all-weather road' had been completed. For reasons of economy the preservation society decided to reopen the line as a 2-foot-gauge railway. The re-opening, from 1983, as the South Tynedale

Railway, has been successful, but the remoteness is still a difficulty: 'I'm one of the closer volunteers,' an official of the line told the *Railway Magazine*. 'I live 32 miles away.')

Suddenly my guide extended his arm through his window and began jabbing his finger towards a driveway. With mutual horn-tooting we separated as he continued on, and I turned into the station car park. The snow had stopped, but the air was misty and freezing.

To get from the car park to the station you pass the engine shed and railway yard, where there's an engine men's mess. I was once sitting in it at 6 a.m., having a safety briefing from an ex-steam engine driver on BR called Clive Groome, who was about to give me a day's instruction on the footplate. To evoke Clive Groome, it is merely necessary to say 'engine driver', which surely suggests a tall, slender man with a moustache, a sort of Edwardian cowboy. His question to me on that early morning, which was also misty (and would be mistier still after a few hours of my inept handling of the loco), was, 'Do you like your tea strong?' – the implication being that any aspirant train driver *should* like it strong.

In the booking office I stood in front of the fire thawing out for a minute, which was all that was needed, since it seemed almost frighteningly incandescent, being made of Welsh anthracite, no seething lump smaller than a loaf of bread. As with any Victorian station – the theme of Sheffield Park is LBSCR in the 1880s – the ticket window was a case of shouting through what's called a 'pigeonhole'. These little windows were always small, for reasons of security: the railway company didn't want you reaching in and grabbing the money.

A man in an LBSCR cap, who inhabited that little

world made almost entirely of wood and brass, sold me a day rover ticket. It was an Edmondson: a ticket of the design created by Thomas Edmondson in 1836, while he was a station-master at Brampton on the Newcastle–Carlisle Railway. Edmondsons are made of stiff card and look so elemental you would have thought they were of simple dimensions: 2 inches by 1, say, but they are 2¼ by $^{17}/_{32}$ to deter forgers. Previously, tickets had been written out by hand, whereas an Edmondson was printed in advance, subject to the addition of a stamped date (and Edmondson invented the date stamps as well). Unscrupulous tickets clerks might sell the old sort of tickets, then pocket the money, a practice that bothered Edmondson who was a strict Quaker. His tickets had a serial number written horizontally down the side, therefore, so that every sale had to be accounted for.

All the Big Four had printing works for making Edmondsons. The one belonging to the LNER was in York. Under BR, printing was centralised at Crewe. As the rate of line and station closures gathered pace, it was common to see Edmondsons with the station names printed off-centre, one element having been deleted: in his book *Thomas Edmondson and his Tickets*, Michael Farr includes a picture of a platform ticket on which 'Barnstaple' is way over to the left. This is because the word 'Junction' has been removed from the right-hand side. In 1970, when the line from Barnstaple Junction to Ilfracombe was closed, Barnstaple Junction forfeited the second part of its name.

Flamboyant travellers might tuck an Edmondson into their hat bands. The first wallets I owned used to have slits to accommodate Edmondsons, which made me feel grown-up, but these little tickets had gone from the national network by 1990. They tend to disappear in modern wallets, where all the pockets are debit-card-sized. I felt guilty about paying with a debit card at Sheffield Park, because I'd forced the clerk to 'break character' by producing the card machine.

Preservation has kept Edmondson's name alive, since almost all the lines use them. One supplier of Edmondsons to the early preservationists was the Eton College Press. Another was Colin Burdett Wilson of Cheltenham, author of a book on Great Western advertising material. According to Michael Farr, 'He printed Edmondson tickets for several preserved and miniature railways in which he had an interest, including the Great Western Society, Dean Forest Railway and some special events such as annual dinners of railway societies.'

In 1987, the Transport Ticket Society persuaded the above-mentioned South Tynedale Railway – the nearest preserved line to Brampton – to name one of its tank engines *Thomas Edmondson.*

The Bluebell prints Edmondsons from old presses, both for itself and other lines. The West Somerset, Isle of Wight, Severn Valley and Swanage Railways do the same. I have a small collection of preserved-line Edmondsons – not a formal collection; merely a scattering in my desk drawer – and none of them are clipped, as an old Edmondson would have been unless I had evaded the ticket inspector. The tickets could be clipped in various ways to show a journey completed in part or whole, but this need doesn't usually arise on preserved lines, where the ticket is simply a pass to ride up and down until the railway closes for the evening. Edmondsons in their heyday were quite brutally treated. A child's ticket, for example, might be made by snipping an adult one in half, and it would be a very contemptuous snip if the ticket clerk thought the buyer was lying about their age.

On Platform 1 at Sheffield Park, there seemed to be more station-masters than members of the public, or at least more Bluebell officials, who tended to be men of retirement age in long black coats with LBSCR insignia. Where did they get their regalia? A website known as HOPS was mentioned. It stands for Heritage Operations Processing System, and it offers web-based programs for administering preserved railways. Clickable options include 'Operations calendar', 'Rostering of staff', 'Competence management', 'Short-notice cover'. There is also, less intimidatingly, an online shop. A Southern Railway cloth badge reading 'Signalman' is £1.50. It is a reproduction, I think. A reproduction LMS waistcoat and tie costs £38.77. The waistcoat has metal buttons. 'To complement your waistcoat and keep you safe in an operational environment, your waistcoat comes complete with a clip-on tie in maroon ... You will look the part and be safe as a Guard, Porter, Signalman or Station Master on duty at your preserved railway.' Ticket clippers start at £17.99. A riveted leather driver's bag is £100.

The talent of the preserved railways for finding anniversaries was evident in that my visit coincided with the Bluebell's commemoration of the deliberately bad 'Sulky Service' laid on by BR, which had ceased sixty years ago to the day. 'Sulky Service' trains would have made their grudging stop at Sheffield Park having previously made an equally grudging stop at the station to the south, Newick.

But today, Sheffield Park is the southernmost station and Newick has been demolished (although the Bluebell does not rule out trying to reach the site again). In order to mimic an arrival from Newick, one of the station-masters told me, trains would be running into Sheffield Park from a siding.

'But how,' I wondered, 'will the punters know that the train has come from Newick, or is *pretending* to come from Newick?'

'Because,' came the answer, 'the announcer will say, "The train now arriving *from Newi*ck is the 11.30 for East Grinstead," or whatever. But the whole plan's up the spout because the S15's not working.'

The S15 class of engines were built by the London & South Western Railway in the 1920s and 1930s – too early for the class to have a memorable name. Seven were rescued from Barry scrapyard for preservation. 'It's a straightforward, strongly built engine,' another Bluebell chap (this one covered in oil) told me; but it wasn't working today.*

As it turned out, I spent all morning at Sheffield Park, which is perfectly easy to do. It's a rangy place, like a little railway colony, with pub-cum-restaurant, bookshop, engine shed and, on Platform 2, a museum. My favourite exhibit here was a glass-fronted display of Southern Railway holiday brochures, featuring illustrations of women in cloche hats and men in Gatsby caps lounging about on the Sussex Coast or on the Isle of Wight. Even more exciting were the brochures for the boat trains,

*It might be asked why they couldn't just fire up another engine. But that takes at least four hours.

including the Southern's guide to Picardy and Paris, and (for the French market) *Outre-Manche et ses Plages*, with a picture of Brighton looking semi-tropical. It was almost as good as standing by the fire in the booking hall. One of the museum's curators (there were at least two) told me that *The Titfield Thunderbolt* was being shown on a loop somewhere in the station. He himself wouldn't be watching it ('I know it off by heart'), and I knew it sufficiently well not to need to seek it out. None of the curators in the museum was able to tell me anything about Margery Bessemer (after whom the station pub, the Bessemer Arms, is named). She seemed to have made that one decisive intervention, creating the breathing space for the preservation society to form, before retreating back into the shadows.

Three Engines

I lunched in the pub and, when I stepped out again, an engine – a comparative rarity on that day – was rolling into the station with some very green carriages behind. The engine was a Q Class, No. 30541– a modest-looking goods engine built by the Southern Railway and used by BR until being sent to Barry scrapyard. It was liveried in BR black, and the driver was a woman, impeccably made up and looking regal with one hand resting lightly on the regulator. I think I recognised her as being Liz Groome, daughter of the above-mentioned Clive, but he has three daughters, all of whom fire and drive engines, so I couldn't be sure.

Clive Groome had been in my thoughts all morning, partly because of the engine that stood in a siding at Sheffield Park. This was a small tank engine, specifically a London Brighton & South Coast Railway 0–6–0 T Class A1, numbered 672 and called *Fenchurch*. It was built in 1872 by William Stroudley, one of a successful class of engines designed for rapid acceleration between close-together stations on the LBSCR's South London line – so *Fenchurch* would have been a familiar site to all those Mr Pooters waiting at places like Clapham Junction, Denmark Hill, Peckham Rye. Later, *Fenchurch* was sold to the Newhaven Harbour Company, who wanted *Fenchurch* for its lightness: it was able to traverse the swing bridge between the East and West Quays. The A1s were nick-named Terriers because their exhaust beat sounded like a bark, and a Terrier spotted anywhere was usually on its way to or from some delicate piece of track. *Fenchurch* remained on active service at Newhaven under BR and was for a while the oldest of BR's engines. The Bluebell bought it in 1964.

The course I did with Clive Groome was spent on the Bluebell's other Terrier, *Stepney*, which was built in 1875, and was bought by the Bluebell in 1960, its very first engine. After its South London duties for the LBSCR, *Stepney* worked the Hayling Island branch in Hampshire, where it specialised in not destroying the bridge over Langstone Harbour.

Stepney is today the unofficial mascot of the Bluebell, and the engine most closely associated with the line, mainly because of the Reverend Awdry's book of 1963, *Stepney the Bluebell Engine*. The Foreword describes how Percy, 'a kind-hearted little engine', is distressed 'because many fine steam engines are cut up on the Other Railway (BR)'. But Percy's ideas are 'a little muddled. British Railways Officials are *not* cruel. They are ... glad to help engines go to places like the Bluebell Railway at Sheffield Park in Sussex'. In the book, Stepney visits the Island of Sodor from the Bluebell for a sort of working holiday, which is a great compliment to the Bluebell, because while Awdry had volunteered on the Talyllyn in the 1950s, that line would not make it into the Railway Series until much later.

Fenchurch is liveried in dark brown, which is apparently 'the Bluebell version of the Newhaven Harbour livery', whereas *Stepney* is liveried in a colour of William Stroudley's invention called 'Improved Engine Green', which has caused a lot of fraught discussion on railway forums, because Improved Engine Green is mustard yellow ... or orange ... or gambonge, depending on the eye of the beholder. It was been suggested that Stroudley was colour-blind. Even so, he is considered one of the 'Artist Engineers', and *Stepney* is a very pleasing object that would look quite appropriate, if miniaturised, at the base of a Christmas tree with a ribbon round it.

In 1961, the year after its arrival at the Bluebell, *Stepney* appeared in a BBC TV version of *Anna Karenina*, starring Anna Bloom and Sean Connery. I bought the DVD, being interested to see whether little *Stepney* – known for his kindness to bridges – could possibly have been the engine that Anna leaps in front of at the end. I think it *is Stepney* in the fatal scene, and then there is a quick cut (rather lurid) to *Stepney*'s underframe, followed by a shot of a steam-shrouded gas lamp sputtering out.

I did not know the history of the engine when I was on its footplate. If I had done, I might have been even jerkier in my handling of the regulator. I did improve by the end of the day, because Clive Groome is a good teacher. He is always impeccable in overalls and hobnail boots (no Doc Martens on the footplate: the plastic soles melt) with a spotted handkerchief worn around his neck. In his mid-eighties, he could pass for late-fifties. His instructions are concise and orthodox. He speaks of 'preparing the fire bed', 'trimming the coal', 'giving a pop on the whistle'. He became a 'passed' or qualified fireman in 1951. In 1954, he was firing on the 'top link' (expresses) from Nine Elms engine shed in South London. He became a driver in 1961. And then there is this, from his website: 'Between 1967 and 1979 he was trained on a variety of diesels and electric machines, but was unable to maintain his interest in a job transformed by modernisation and de-skilling. As a result he resigned in 1979.' Clive now teaches under the strapline of Footplate Days and Ways. Besides the Bluebell, he can be found at the short Lavender Line (between Uckfield and Lewes) and the Llangollen Railway in Wales.

Horsted Keynes

Liz Groome – or let's say 'Ms Groome' to be on the safe side – had now 'run round' her train in the Q Class, and she was ready to leave for Horsted Keynes. Her train consisted of BR Mark 1 coaches, and earlier coaches designed by Oliver Bulleid, Chief Mechanical Engineer of the Southern Railway from 1937 and then of the Southern Region of BR until 1949. The carriages built to his designs for the Southern (and retained by BR) are pleasingly curvaceous. They are spacious, bulging over the tracks and making full use of the loading gauge. They have wooden bodies but with steel sheeting, and the livery is dark green: Malachite Green, which was Bulleid's green, whereas his predecessor, Richard Maunsell, had favoured

Olive Green. These Maunsell carriages form the core of the Bluebell fleet, and the Railway considers the Bulleid open thirds ideal for its purposes with their 'high seating capacity and wide windows ... The deeply sprung and padded seats in these coaches are probably the most comfortable ever provided for ordinary passengers.' That said, I have never seen an old third-class coach that was not more comfortable than any modern standard class carriage. Fire regulations deprive us of the wooden interiors that featured on the Bulleid coaches, also the seating thickly padded with horse hair; and modern corporate colour sense deprives us of the romantic, autumnal hues of the old seat moquettes.

As we pulled away, I was looking through the window. None of the trademark wild bluebells yet, in the frosty fields; just one typically self-assured fox. I then fell asleep – a tribute to Mr Bulleid – for the remainder of the ten-minute journey.

Horsted Keynes has a 1920s theme, which is why it played the role of Downton Station in *Downton Abbey*. It is painted in Southern Railway station colours – two shades of green and cream. I felt it would be churlish not to go on a tour of the Bluebell carriage works, which is at Horsted Keynes. The place was superbly restful: the sweet smell of wood being worked, and the mellowness of mood that always seems to accompany that activity. A radio played quiet jazz; a cat flitted about.

Our guide (far better spoken than anyone likely to have been found in a carriage works of the past) pointed to a wooden skeleton. This was a carriage built to the design of William Stroudley, the Terrier man. It had been rescued from 'farmyard use'. He indicated a varnished teak Metropolitan Railway bogie stock carriage dating from 1900. It had been running on the country end of the Metropolitan Line – shuttling up to Chesham behind steam engines – until 1962, and it's depressing to think that I was alive (albeit only just) when this was going on. The Bluebell – desperate for rolling stock in its early days – bought four of these carriages because they were cheap: £60 each. They were used in the early years, but then put out to grass, as it were, in the Horsted Keynes shed, making occasional appearances for filming – including in John Betjeman's documentary of 1973, *Metroland*. They were restored at the turn of the twentieth and twenty-first centuries, and in 2007, the Bluebell won the Heritage Rail Association Carriage and Wagon Award for the work.

Somebody on the tour (there were four of us) asked whether any carriages on the Bluebell were gaslit.

The answer: 'No, but the West Somerset Railway is restoring an old Great Western sleeper carriage, and that *will* have gas.'

The volunteers at work tended to be doing their own thing very meditatively with paint brush or sand paper, and the mood of absorption was summed up in an obituary I read on the Bluebell website when I got back home. It concerned a volunteer called Chris.

> Well known in the C&W department for painting carriage ceilings, taking advantage of his considerable height, Chris was also an important part of the Horsted Keynes Pullman Car Fund, preferring to be involved quietly behind the scenes. Most recently he had taken home a large proportion of the interior of the Bulleid composite No. 5768, which is currently

> being restored, a few bits at a time. Each piece of timber would return a week or two later resplendent after numerous coats of varnish.

Readers were invited to make a donation in his name to the Railway.

It was now dark as well as freezing, and I was too late to take a connection for East Grinstead. Instead, I took a seat in the waiting room, which I had to myself, and where one of those preserved railway fires was burning so brightly as to be hard to look at. In spite of the cold, the mood was colonial, with an aspidistra on the table, green stained-glass windows, and old suitcases and trunks stacked up for scenic effect – and the ripe banana that I now began to eat while finishing off my flask of tea. I thought, as I often do on preserved railways, that you don't really need the trains at all. But I waited for the last one back to Sheffield Park, while reading a Bluebell publication about the triumphant push to East Grinstead.

The Bluebellers had always wanted to regain East Grinstead, and by 2003 they had purchased the various plots of privately-owned land along which the line had once run. BR donated to the Railway the Imberhorne Viaduct south of East Grinstead, but a cutting just south of the viaduct had been filled with 3.4 million cubic feet of rubbish by the local council in the 1970s, as though to add insult to the injury of the line closure. The good news? The rubbish was not toxic. From 2008, the Railway began removing the rubbish, initially with lorries but then – as though to confirm their possession of the moral high ground – by the more environmentally friendly mode of rail; much of the waste would be recycled. Money was raised through a public appeal, an appeal to members (a 'Tenner for the Tip') and a share issue; there was also a donation from the Mid-Sussex District Council.

The Railway's shops sell a DVD called *Returned to East Grinstead.* Footage shows giant modern diggers in the cutting: it almost looks like sci-fi, and this was, in the words of the narrator, 'the toughest extension of any preserved railway'. The total cost of progressing from Horsted Keynes to East Grinstead was in the region of £30m. On 7 March 2013, the final track bolts were tightened by Barbara Watkins (known on the Bluebell as 'Mum'), and this was the beginning of a series of those things at which preserved railways excel: celebrations. On 23 March, a cavalcade of Bluebell stars, including of course little *Stepney*, rolled into the new Bluebell station at East Grinstead, which is really just platform 3 of the modern station, so there is an easy connection to the modern world, and, as though to prove the point, a special excursion ran a week later from Victoria Station to Sheffield Park. The film shows 'Sheffield Park' on a dot matrix indicator at Victoria, so here was the Bluebell with its name (or at least one of its station names) in lights. The DVD features an interview with the new station-master at the Bluebell's East Grinstead station. 'The line's eleven miles long – forty minutes or so, and that's about right.' If it were any longer, he explained, people might lose sight of the uniqueness of the experience. They'd start reading books or (more likely) checking their phones. 'It mustn't become just another railway journey.'

The Keighley & Worth Valley Railway

Cryer Meadows

The night before I went to the Keighley & Worth Valley Railway, I spent in the Midland Hotel, Bradford, which had been built by the Midland Railway in 1890. The Midland was famous for its railway hotels, and my stay was quite luxurious, doubly so, since this was a Sunday evening, and I seemed to have the place to myself. The Midland goes in for a bit of preserved railway theatrics, in that the sloping walkway leading from the hotel towards Bradford Forster Square Station is punctuated with dusty old trunks and suitcases. The suggestion is that this is still 1890, but the sight of the modern Bradford Forster Square station – which is no longer at the foot of the walkway but lies some way beyond – soon corrects that thought.

In 1890, the station was capacious, with an overall glass roof and a covered forecourt that those emerging from the hotel found themselves underneath. In the early 1960s, this was replaced by a more open-air structure, with the platforms protected from the elements by 'butterfly awnings' that had a certain brutalist glamour. In 1990, the station was rebuilt again, and it is entirely nondescript. As though to point up the bathos, an arcaded wall of the old station remains standing to the east of the new one. The station serves two lines that are at once intensively used commuter lines and pretty rural through-routes: the Wharfedale Line to Ilkley and the Airedale Line to Skipton. Beeching, misreading his crystal ball as usual, had wanted to close most of the stations involved, and many did close, and some later re-opened. Today, it is still possible to go from Bradford to Keighley, starting point of the Keighley & Worth Valley Railway, on the Airedale

Line, although the journey is a rather equivocal experience. The Class 333 electric multiple unit doesn't seem to know whether it's a train or a tram, and the line is half pretty rural branch, half intensively used commuter route.

At Haworth Station on the Keighley & Worth Valley Railway, a more emphatic statement was being made: here is the headquarters and fulcrum of a pretty rural branch line, preserved as it was, well ... in the *past*. 'We're broadly Fifties,' an official of this most relaxed of preserved lines told me. The six stations of the 5-mile line are maroon and cream, the colour (more or less) of BR London Midland Region stations; also of its Big Four forerunner, the London, Midland & Scottish Railway, and this maroon takes its place quite naturally alongside the greenery of the valley the line occupies and the grey of the drystone walls and the tall houses. You'd think the line had always been in the valley, co-existing with the Bridgehouse Beck, which frolics on either side of the line – and the historical continuum is smooth, because the KWVR has the whole of the old branch. The KWVR is unusual in possessing an entire branch together with the connection to the main line – at Keighley. The Ecclesbourne Valley Railway in Derbyshire does run trains over the entire Wirksworth branch, with a main-line connection at Duffield, but the track is leased.

On this bright, slightly smoky morning, a white-haired man who looked like the Old Gentleman in *The Railway Children* walked onto the station forecourt to set down a blackboard reading 'STEAM TRAINS TODAY'. It was quite a modest pitch, compared to what was going on up the hill, where the Haworth Parsonage, Mecca of Brontëland, is located.

A certain John McLandsborough visited Haworth

in 1861, 'a pilgrim at the shrine of Charlotte Brontë', which, considering he was a sewerage engineer, suggests an interesting breadth of character. He also had experience of railway-building, and he proposed a branch running south-west from Keighley to Oxenhope, taking in Haworth on the way. Keighley station had been opened in 1847 on what was then called the Leeds and Bradford Extension Railway. That was the year in which *Wuthering Heights* was published, and the year before its author, Emily Brontë, died.

In August 1847, Charlotte Brontë had arranged for the manuscript of *Jane Eyre* to be despatched to her London publishers, Messrs Smith and Elder, by rail from Keighley. It is possible she took it to Keighley Station herself either by horse and trap or on foot. She had certainly ascertained that the station could not accept money for the despatch of a parcel. The cost had to be paid by the receiver, so in her accompanying note, Charlotte was forced to write, 'If, when you acknowledge receipt of the MS you will have the goodness to mention the amount charged on delivery – I will immediately transmit it in postage stamps.' In her notes to *The Letters of Charlotte Brontë, Volume One (1829–47)*, Margaret Smith writes that what Charlotte described as the 'small station house' at Keighley was 'A temporary structure ... perhaps like that at Bingley on the same line, where, in Nov. 1849, the clerks' room was a wretched wooden hut "about three yards square", and where "during wet weather the clerks transact business with an umbrella over their heads."' But if railways were slower in those days, publishers were quicker. George Smith read the manuscript within a week and published it six weeks later.

The new branch line came too late to benefit the

Brontës, unless we count their posthumous reputations. The purpose of the line was to serve a dozen woollen mills in the valley. They needed the railway to bring in better-quality wool than could be obtained from sheep grazing on the sparse Pennine hills.

The line opened in pouring rain on 13 April 1867. There was a banquet in the Mechanics' Institute in Haworth. Tourism helped its revenues as the reputation of the Brontës grew: the Brontë Society ran a special excursion to Haworth in May 1895, and since the Brontë pilgrims tended to be well-heeled, there was an unusually high incidence of first-class accommodation on the trains of the branch.

The branch had been operated from the start by the Midland Railway, which gained full control of it in the 1880s. In 1921, there were about twenty trains a day along the branch, including a busy Sunday service, four signal boxes open seven days a week, and some freight trains running at night. In his excellent guide to the line, *Brontë Steam*, Robin Jones writes, 'There were only two time-tabled freight trains a day but many ad hoc ones pottered around the branch using the loops and yards as refuges to get away from the passenger trains.'

Then decline set in. Keighley Corporation was a keen operator of trolleybuses, then buses – for which there was much demand, since the small towns or villages served by the branch tended to be at the tops of the hills, with the stations at the bottom. (Haworth Station is only really 'convenient' for Brontë Parsonage if you're quite fit.) After the Second World War, the mill looms were electrically operated, which meant the branch lost much of the revenue it had earned from bringing in coal.

Sunday trains were stopped in 1947; by the 1950s, the line was making a loss. The last passenger train departed from Keighley on the night of 30 December 1961. 'Snow blanketed the valley on the last day,' writes Robin Jones, 'creating a white landscape of the kind that had inspired the Brontë sisters ... The last train left Keighley at 11.15 p.m., with the mayor and mayoress of Keighley, the Keighley station-master and around 150 other passengers on board.' He adds, 'That evening, the Platform 4 sign at Keighley, indicating the Worth Valley arrival platform, crashed to the ground for no discernible reason.'

The last train was a four-car DMU. But it *wasn't* the last train, and not even the last BR train, because on 23 June 1962 BR agreed to run a steam special, and they did so at the bidding of the Keighley & Worth Valley Preservation Society.

Robin Jones differentiates the KWVR movement from those that preserved the Middleton and the Bluebell. The line was saved by local residents, not rail enthusiasts, and their aim was not to preserve steam, but to preserve a service. In this respect, 'what happened on the Worth Valley was far more true to the Ealing Comedy [*The Titfield Thunderbolt*] than its inspiration, the Talyllyn, had been.'

The Keighley & Worth Valley Preservation Society was formed at a meeting held in the Keighley Temperance Hall on 24 January 1962. It had 266 members, and Bob Cryer was elected chairman. Cryer had been born in Bradford; he graduated from Hull University and became a lecturer. He died in a car crash in 1994, by which time he was the Labour MP for Bradford South, having previously been the MP for Keighley. Before coming to the KWVR I had spoken to his widow, Ann Cryer, who herself was MP for Keighley between 1997 and 2010 and is the current chair of the KWVR.

'Bob was a socialist and a co-operator,' she said, 'and he wanted to preserve the railway for the local people, but he was also a rail enthusiast. As a boy, he'd lived at 15 Albert Road, Saltaire, and he and his little chums would climb up on a wall at the end of the road and take the numbers of the trains on the line between Shipley and Bingley.' She married Bob in 1963; their honeymoon was on the narrow-gauge railways of Wales. They drove there in his two-tone-green Armstrong Siddeley. Like L. T. C. Rolt, Bob Cryer combined rail and car enthusiasm; also like Rolt, he wanted a democratic organisation for his preserved line: 'The Society,' Anne told me, 'would have the control and the shares; there would be no Fat Controller.' The aim was to run a combination of heritage steam trains and timetabled diesels, for the more practical needs of the locals.

It would take six years before the Society reopened the line. At first, the Society hoped to lease it from BR, as the Bluebell people had done with theirs. But BR demanded the Society buy it outright – for £34,000, which was almost a giveaway, being the cost of about ten average-sized houses. It was cheap because a number of bridges were in need of repair. But it was money the Society did not have. In 1963, a provisional agreement was reached with BR: the Society would buy the branch on hire purchase over twenty-five years, and from January 1965 the Society began maintaining the branch. In 1966, a company called the Keighley & Worth Valley Light

Railway Limited was formed, with most of the shares owned by the Preservation Society.

The first engine had already arrived: a very small tank engine, ex of the Lancashire & Yorkshire Railway, of a type nicknamed 'Pugs'. It was appropriate that this should be the first engine on the KWVR, because kits by which model Pugs could be made represented the entry level for Airfix modellers at the time. Mindful of their use in industry and freight haulage, modellers would 'distress' the locos with hard-water staining and dirt colouring, but the Pug that came to the KWVR would be pulling passenger coaches.

The grand reopening took place on 29 June 1968. Seeking a celebrity to cut the ribbon, the Railway approached Dr Beeching, who turned them down, so the Mayor of Keighley was called in. BBC and ITN newscasters were present at the re-opening, preserved railways still being a novelty. Robin Jones gives an exuberant ten full-colour pages to the event, listing the Top 30 of the UK singles chart in that week: Gary Puckett and the Union Gap were at Number 1 with a number that could not be written today, 'Young Girl' ('My love for you is way out of line'). Motive power for the train was provided by two tank engines. The first was an Ivatt 2–6–2, 41241, which was built by BR to an LMS design. When working for BR it was painted black, but the KWVR – not as uptight about liveries as some – painted it in a version of Midland maroon with 'KWVR' written in yellow on the sides. The other tank engine was an American shunter or 'switcher'. It'd had various liveries in its time, but the KWVR painted it golden brown. The train ran along the line from Keighley, cheered by big crowds all the way. At

Oxenhope, the Whitworth Vale and Healey Prize Band struck up with 'Congratulations' by Cliff Richard. Robin Jones, forgivably carried away, writes, 'It was evident to all that the miracle had been achieved, one that almost certainly would have inspired the Brontë sisters had they been alive at the time.'

We now come to some other literary sisters – two of them – and a brother: Roberta, Phyllis and Peter. E. Nesbit's novel, *The Railway Children,* helped the KWVR twice over. A TV version was filmed on the line in early 1968 (before the reopening), and the first episode was broadcast in May 1968, immediately *prior* to the reopening. The actor Lionel Jeffries was persuaded by his daughter – who had loved the TV series – to buy the film rights.

'The railway didn't really have a phone,' Ann Cryer recalls. 'I mean, its phone number was the number of our house at Providence Lane, Oakworth. It was a two-tone green trimphone – we were very keen on green – and it was always kept on the box that held my Singer sewing machine. One evening towards the end of 1969, it rang, and it was a chap called Bob Lynn. He said he was a TV producer, and asked if I'd heard of Lionel Jeffries. He said, "Lionel Jeffries wants to make a feature film of *The Railway Children* on your line."

'I said, "Is this somebody having a joke?"'

The film was released on 20 December 1970, which date might be said to mark the *real* launch of the Keighley & Worth Valley Railway. It was the making of the line: ticket sales doubled overnight, and the film helped preserved railways in general, alerting the public to the fact that there were a number of railway film shoots (as it were) permanently ongoing, in which anyone could have a walk-on part. As cultural creations, only Thomas the Tank Engine and The Titfield Thunderbolt have done more to promote preserved railways, but it should be added – by way of coming down to earth – that the famous KWVR has never forgotten its socialist roots. It has always timed its services to be helpful to local people, and it has recently offered local residents special railcards, giving a discount on the tourist fare.

Diversion to Barry

At Haworth Station, the Old Gentleman sold me a ticket to go up and down the line, and I waited on the platform, alongside a couple of happy families with toddlers. A bell rang, signifying the imminent arrival of a train, as even the toddlers seemed to know (judging by their excitement), although modern trains are no longer announced by the ringing of a bell. The engine of the day, Midland 4F 0–6–0, 43924, came around the bend in its portable cloud, dragging its rake of maroon BR Mark 1 coaches, and it clearly did not disappoint the toddlers. But the engine was even more distinguished than they knew, because 43924 was the first engine to escape from Barry scrapyard, a feat that requires a diversion to Wales.

Officially, this scrapyard in south Wales was called (and owned by) Woodham Brothers. Dai Woodham – a handsome, amusing man with a social conscience – was a scrap merchant or 'junky', as he called himself. He had entered the family business in 1947 after a good war, in which he had won the British Empire Medal. When he first heard about the 1955 Modernisation Plan, with its aspiration to get rid of 16,000 steam locos and 400,000 wagons, he thought, 'Here's a wonderful gravy train, and I'm going to get on it.' It took three years for him to get on it, since BR began by scrapping its engines in-house – often at the places where they had been made, so there were 'steam dumps' at the sites of the great engine sheds: Swindon, Darlington, Eastleigh, Doncaster and Crewe. (In the days of the London & North Western Railway, two locos a week were made at Crewe: in the late 1950s, twenty a month were being cut up there.) From 1958, when the programme was intensified, the scrapping was outsourced to private scrap merchants, and Dai Woodham tendered for business. His deal was unusual, in that both wagons and locos would be sent to him for scrapping. In 1959, Woodham went to Swindon Locomotive Works to learn how to scrap locomotives. 'I got drunk every night with the foreman, who was a Scotsman,' he recalled.

The engines began to arrive at Barry, some under their own steam, others in convoys of the condemned, sometimes hauled by a steam engine that would also not be returning, or by a smug diesel that *would* be – conjuring an image that seems like some nightmare in the mind of Reverend Awdry, an X-rated Thomas story. Woodham's scrapyard was in Barry Docks, and he had rented some adjacent disused railway sidings to store the locos. So the docks were becoming an industrial graveyard twice over, in that those sidings had been made available by the decline of coal exports. In 1958 Woodham scrapped twenty-eight locos. The next year, he scrapped nine. In 1960, he scrapped only one, and 1961 – when Woodham cut up twenty-one locomotives – would prove to be the last year of significant scrapping. Scrapping at Barry stopped entirely in 1965, even though locomotives would continue to arrive until 1968.

This apparent timorousness is explained by Woodham's policy of scrapping wagons first. They were easier to scrap, so there was a quicker turnover of cash. Woodham said he was saving the engines for a 'rainy day', of which, this being Wales, there were many – and then there was the corrosive salt air. Here were a couple of hundred locos assembled in one place – a gricer's dream come true – but they were all rusting corpses, some with their smokeboxes ajar like opened tin cans. But if Barry Docks was a hell for steam locos, Satan turned out to be a nice guy.

In the late 1960s, it became apparent that the preservationists wanted to buy the locos at Barry, which suited Woodham because then he could get money for them without going to the trouble of cutting them up. But if his motivation was primarily commercial, he was also genuinely sympathetic to the preservationists.

The first engine to escape from Barry was the one we have just glimpsed enjoying its retirement on the KWVR: 43924. It was bought for £2,000 by the Midland 4F Preservation Society. On 10 September 1968, they towed it to Yorkshire behind a diesel traveling at 15 mph. They'd spent many hours lubricating No. 43924, because BR had warned them that if it seized up on the main line, it would be cut up. The engine was in steam again in 1970, and was sold to the KWVR in 1987.

Two engines escaped from Barry in 1969, eight in 1970, ten in 1972. Preservationists went to Barry to browse, then to earmark an engine by putting down a deposit. Warnings were scrawled in chalk: 'Boiler reserved – Mid-Hants'; 'Bought by NYMR. Take no parts.' Parts *would* be taken. Big ones – coupling rods, buffers, safety valves, chimneys – were likely to have been taken maliciously by rival preservationists, but Woodham didn't think *mens rea* was always involved. 'Police would call me, and say they'd arrested some man who would usually have a wonderful job – a town clerk or something – and he'd taken something like a stop valve off an engine. But I would never prosecute. He wasn't stealing, he was taking a souvenir, and after a while the police stopped prosecuting.'

The purchasers would come to Barry to paint or de-rust the engines on which they'd made a down payment, and Woodham noticed, 'Many women came to work on those engines.' It might take years for a railway or syndicate to raise the purchase price – throughout the 1970s, anyone on a railway mailing list would be bombarded with appeals for assistance – but Woodham would never allow gazumping.

The scrapyard became a tourist attraction, and old films and photographs suggest the staple visitor was not so much a man in an anorak as a man in a rally jacket – a racier type, long-haired and with a cocky demeanour. One such was Nicholas Whittaker, author of *Platform Souls*, the best book about trainspotting. He visited Barry with his mates in 1969.

> This was a weird landscape, painted every imaginable shade of decay. Some of what we saw was disgusting: strange formations of fungi and rust, the cancerous growths that had taken down our lovely steamers ... A kind of demob fever took hold of us ... we could do whatever we wanted: climb up on the cab roofs and walk along the top of the boiler, stick our heads down

a chimney and shout four-letter words into a loco's iron belly.

This was his revenge for years of being chased off railway premises. Back home in Burton-on-Trent, Whittaker took out his steam books, and marked off the engines he'd seen at Barry: 'There were lots of cops: Castles and Halls from the Western, Battle of Britains and Merchant Navys from the Southern ... but I couldn't feel the same joy I'd felt before. It felt more like a duty that had to be done, that was all – my final audit of the steam age.'

But the *preserved* steam age was just getting going. In 1974, nineteen engines were removed from Barry, one of which was 71000 *Duke of Gloucester*, a Pacific, the sole member of Standard Class 8 and sometimes referred to as the only truly express passenger locomotive to have been built by BR. In *Lost Railways*, Peter Herring describes 71000 as 'a lost orphan prevented from fulfilling its potential'. It was fitted with a Caprotti valve gear – a relative novelty, regarded as more efficient than a Walschaerts or Stephenson valve gear. For those who can understand it, I also offer this from Herring: 'The intention was to marry the Caprotti gear to the Kylchap exhaust system, unquestionably the finest device of its kind. To the dismay of the design team, however, an orthodox double chimney and blast pipe were installed.' As a result, the engine had steamed badly, hence its short operational life: 1954–62. It was rescued by the Duke of Gloucester Steam Locomotive Trust, who took it to Loughborough on the Great Central Railway by lorry. (From the mid-1970s, most removals were done by road, and Barry became a mecca for another sort of hobbyist: appreciators of big-scale road haulage: the movement – usually done by night, ideally with police outriders – of 'abnormal loads'.)

There is footage on YouTube of 71000 being unloaded from the back of a lorry at Loughborough. It seems the whole of the town has turned out to greet it, including a young girl flitting about in a poncho. Suddenly she darts towards the engine and puts a hand on its buffer beam, to assist the engine's removal from the low-loader – a token effort, or perhaps the bestowal of a blessing. The rebuilding of *Duke of Gloucester* (complete with Kylchap exhaust) was finished in 1986.

In 1979, the Barry Locomotive Action Group was formed to help facilitate further removals of locomotives from Barry. The last removal was in 1988. Of the 297 locomotives that were taken there, 213 were saved. Most of the Barry engines came from the Western and Southern regions of BR, but they went from Barry to all parts, and in complicated ways. Here is a typical entry in a book called *Barry Scrapyard* by Keith W. Platt. It refers to a GWR 0–6–0 tank engine:

> No. 9629 was purchased for Holiday Inn for display outside their new hotel in Cardiff. It left Barry in 1981 for cosmetic restoration at Carnforth and was displayed in Cardiff from 1986. The Marriott group obtained Holiday Inn and they donated the loco to the Pontypool & Blaenavon Railway in 1995. A group have been involved with its restoration ever since then, including its boiler, which had been sold to another group.

Today about 80 per cent of the engines on our preserved railways are ex-Barry inmates. Dai Woodham received the MBE in 1987, for the promotion of industry in Barry; he died in 1994.

Along the Keighley & Worth Valley Railway

We will begin where the line begins, at Keighley: a countrified town of castle-like old mills.

The Keighley & Worth Valley Railway describes Keighley Station as its 'main connection with the outside world', and I admire their bravery in parading their dreamworld in so public a setting, like a man I know who wears Georgian clothes all the time. There are four platforms in all at Keighley. Of the two serving the national network, one is for trains to Skipton, the other is for Leeds or Bradford. The two KWVR platforms might seem almost normal to some onlookers, until a steam train pulls in. Red enamel signs reading 'Telephones' and 'Refreshments' are subtly of the past rather than antiquated. While the KWVR owns the whole of its line, these two platforms are leased from Network Rail, and this situation permits some through ticketing between the KWVR and the national network. You can buy a ticket at, say, Haworth for Skipton, or vice versa.

The line climbs from Keighley, making smoke, as the old mills once did. My ticket was politely checked by a young man (this was the school holidays) soon after departure and would be checked several more times by what I first thought was the same young man, but there turned out to be two of them, both called Joseph. One was sixteen, the other fourteen. Both wanted to work on main-line railways. One Joseph lived opposite Ilkley

Station on the national network. The other lived opposite Ingrow West Station, which is the first stop after Keighley, and which looks 150 years old, but was rebuilt in the 1970s, the original having become derelict. On the platform here, the Josephs introduced me to a young woman, Faye, who was eighteen and had just left school. She had been working in customer support at Leeds Station. She wanted to be a train driver on the national network and was a trainee fireman on the KWVR.

The volunteers were all either old or young, much like the typical visitors to preserved lines, who are grandparents with grandchildren in tow. During the stop at Ingrow, a Joseph introduced me to the steward in the buffet car. Ken, presiding over the usual abundance of booze, told me he would be 'seventy-nine next birthday'. He had previously been a steward on BR. Shortly after he retired, he was watching TV in his living room when his wife came in and demanded, 'Remind me: how long have you been retired?'

'Three days,' said Ken.

'Well, you're driving me mad,' his wife said. 'Go out and do something.'

In the yard of Ingrow Station on the KWVR stands 'the Engine Shed', which 'tells the story of the steam locomotive', and is run by the Bahamas Locomotive Society, which recently overhauled the LMS Jubilee Class 5596, *Bahamas*. The yard is also home to the Museum of Rail Travel, run by the Vintage Carriages Trust, which began accumulating wooden-bodied railway carriages in 1965, when a lot of them were being scrapped; and here we will take another detour.

A few weeks after my visit to the KWVR, I was put in touch with the Hon. Sec. of the Trust, Dave Carr, a retired primary school teacher who lives near York. He suggested we pay a visit to the National Railway Museum in York, and it was as if I'd previously walked around the place with blinkers on. Dave pointed out some of the details of carriage finishing, discussing 'scumbling' – the creation of the effect of wood grain, as done on the 'teak' carriages of the London & North Eastern Railway, some of which were steel, but were required to look like teak. He also talked about lining-out – applying the thin lines that provide colour contrast in a livery. This requires great skill: 'painting lines two millimetres thick by hand'.

We admired the yellow 'go-faster' stripes on the Coronation Class *Duchess of Hamilton*. 'Tasty, is that,' said Dave. 'Very tasty.' 'Lining-out' might also encompass the painting of a shadow effect around an engine's number. The small amount of people who can still do lining-out are well known in the preserved railway industry and in constant demand. Such skills are in danger of dying out. 'Try finding someone who can weave luggage-rack nets,' said Dave. 'Actually, the wife of one of our members learned how to do it. You create a knotted net, like those fishing nets that are now illegal because they take the slime off fish.' Every year, there's a convention of carriage restorers. 'It's run by Stephen Middleton,' said Dave. 'He's up at Embsay [the Embsay–Bolton Abbey Railway]. He restored that NER autocoach they have up there.'

Dave Carr became interested in railways when he was at Keighley Grammar School. 'We had an English teacher called Tony Peart. He ran a railway group, and the bloke was charismatic. He used to have permits for shed tours, and if you'd been a bit naughty he wouldn't take you. He was one of the leading authorities on railway lamps, and if people saw him up at the memorabilia auctions they used to have at Cundle's in Malton, they'd be thinking, "What's he seen?"'

As we looked at the salmon-pink seat moquette in an LSWR carriage, Dave mentioned the insipidity of modern colour schemes, and I told him I'd once got talking to a carriage designer (it was on a train actually) who said that any bold colour would be met with objections from some quarter or other, 'So we just say "Let's go 'hospital'"' ... which usually results in pale blue.

When it comes to carriage seats, Dave Carr suggested that 'function had triumphed over form.' Modern, moulded seats may not be pretty, but they can support the body better than the old wood-framed steel-spring-and-horse-hair ones, which have the additional disbenefit of being a fire hazard. On the other hand, carriage seats, unlike people, have not got any bigger over the years. 'The old maxim,' said Dave, 'was sixteen to the bum, sixteen to the ton.' In other words, the average passenger was taken to be 16 inches wide and 10 stone in weight. That's no longer the case.

After Ingrow, the line continues to climb, and there was now a patchwork of tilted fields on either side. As cloud shadows moved across them, I identified several railway greens in the grass – certainly, malachite, Brunswick and olive. Damems Station is claimed by the KWVR as the smallest in Britain, and it does have the dimensions of a garden shed. It's a request stop, and nobody did request it, so I felt sorry for the well-turned out station-master smiling on the platform.

Towards Oakworth, the train is in pure moorland, russet tones overtaking the green. Oakworth is the beginning of the territory covered by the leaflet (available all along the line) advertising *The Railway Children* Walks. It was on Oakworth Station that Bobbie (Jenny Agutter) sees her father emerging from a cloud of steam to such tearjerking effect, and the station cottage was home to Perks the porter (Bernard Cribbins). After Oakworth comes an embankment reverently marked on the leaflet, 'Red petticoat waving'. Bobbie and Phyllis (Sally Thomsett), wave their red petticoats to stop the train, a landslide having occurred. In his novel *Mr Phillips*, John Lanchester writes, 'In Mr Phillips's opinion the sexiest film ever made was *The Railway Children*, although he knows you aren't supposed to say that.' Also very beguiling is Jenny Agutter's confiding narration, beginning, 'We weren't *always* the Railway Children ...'

Next comes Mytholmes Tunnel. As first built, the line incorporated a viaduct over a mill dam here: the trains had to rather tiptoe over it, and some passengers would refuse to remain on board as they did so. In 1892 a tunnel was opened along this stretch, which about eighty years later would facilitate the most harrowing scene in *The Railway Children*: a young runner breaks his leg in the tunnel while following a paper chase; meanwhile a train is due. Whether a paper chase would really have been run through the tunnel of a working railway is less important than the fact that E. Nesbit thought she could get away with writing such a scene. The novel was published in 1906, when a rural branch line was beginning to be seen as complementary to the countryside. Incidentally, the country branch that fired Nesbit's imagination was not in Yorkshire but in Kent, where she had lived during her adolescence, in a big house overlooking the South Eastern Main Line. What seems to have been even more influential was her reading of a novel – *The House by the Railway* – by Ada J. Graves, published in 1904, which has essentially the same plot as her own story. So while Nesbit was thoroughly admirable for her social work, and the luminous quality of her prose (Noel Coward said she had 'an economy of phrase and an unparalleled talent for evoking hot summer days in the English countryside'), she was also a plagiarist.

The railway returns the favour the film did for the line, by encouraging the visitor to see it again, and I watched it when I returned home in the light of what I'd read in the KWVR. You can sort of see that, yes, Sally Thomsett, as Phyllis, is actually older than Jenny Agutter even though playing her younger sister. Her contract precluded her from turning up on the set smoking, or in her sports car, or with her boyfriend.

The main engine used in the film is an ex-Great Western Railway pannier tank, 5775, which also served as a London Underground service engine. The engine was acquired by the KWVR in 1970. Originally the engine had been green; when working on the Underground it was maroon. For the film it was painted caramel, the fictitious livery of the fictitious Great Northern & Southern Railway.

On the approach to Haworth, mills and terraced houses reappear, but always redeemed by nearby grass and trees. On the climb towards the terminus, Oxenhope, one is very aware of the waters of the Bridgehouse Beck, racing the opposite way to the train. There is not much to distract you from the beck, the industrialisation of the valley having faltered by this point.

I was back at Haworth by 5 p.m., and the Old Gentleman was taking in the sign reading 'Steam Trains Today'. There would be steam trains tomorrow, however.

What Happened Next: Overview

The obvious omission from my list of pioneers is the Ffestiniog Railway, which was relaunched (it had stopped operating but never exactly closed) in 1955, but I want to consider it later, alongside its sister line, the Welsh Highland, which re-opened in 1997.

'Once the Bluebell Railway had shown it was possible to reopen a line closed by British Railways,' writes Jonathan Brown in *The Railway Preservation Revolution*, 'the number of preservation proposals grew rapidly. At times it seemed as though hardly a branch line was closed without someone proposing that it should be preserved.' But 'getting new projects to the operating stage, was slow ... by the end of the 1960s the number of operational preserved lines had not reached double figures.' He counts twenty-six 'projects and openings' for the 1960s and 1970s together. He counts another thirty-eight for 1980–2000, and another thirty-two for '2001 onwards'.

The period of fastest growth in preserved lines, then, was the last two decades of the twentieth century. Turnover and traffic increased as the economy generally boomed. The young trainspotters who had established the lines gave more time to them as they reached what was often a well-cushioned retirement. They could afford to be munificent, and in many cases they brought to the railways business acumen, which they applied to the chasing of grants.

Modest grants used to be available from regional tourist boards, but those boards were abolished fifteen years ago. Local Enterprise Partnerships have to some extent filled the breach. Direct government grants are intermittently available, for appropriately situated railways, from the Coastal Communities Fund. Small grants are becoming available from the Arts Council for those few railways accredited as museums (the Isle of Wight Steam Railway is one), or the larger number incorporating museums. The European Union *was* a source of grants, mainly to railways in areas of high unemployment.

Today, the big prize is a Lottery Heritage Fund grant, but match funding (a contribution from the applicant) will usually be required, and the application must be carefully tailored. 'The Lottery Heritage Fund is not primarily interested in the nuts and bolts of railway operation,' one preserved railway official told me. 'It's not enough to say, "We need a new engine." There must be a demonstrable benefit to *people*. How will the money help more people to enjoy their heritage?' This notion of outreach is increasingly shared by the railways themselves: they must bring in new blood, and the pressing need, as Generation Steam dissipates like the stuff after which it is named, is for *young* blood. (See the 'Some Possible Futures' chapter). The North Yorkshire Moors Railway, for example, is using part of a recent National Lottery Heritage Fund grant – £4.6m, the largest one ever given to a preserved railway – to employ a volunteer recruitment officer. An old school in a hamlet near the North Yorkshire Moors Railway's Levisham Station is to become an annex of the railway, to which school groups will be invited. They will do some railway activities and others just for fun, but it is hoped that some of the attendees will find the railway activity the most fun of all and will become involved with it on a longer-term basis. (The railways have to be careful about calling children under sixteen 'volunteers': there is a danger of falling foul of the anti-child labour provisions of the Employment of Women, Young Persons and Children Act 1920, so children are given 'experiences' rather than work to do.)

Some of the Lottery money received by the NYMR will go on replacing three of the thirty-five bridges on the line. Grants generally are needed for big capital projects. Even the most successful preserved railways, with decent operating profits, can't fund these from 'the till', and in this they reflect the big railway, which has always been mired in its capital costs. Most preserved railways have a charitable component in their administrative structures, so as to be eligible for grants or Gift Aid from donations. The operating company itself might be a charity, or it might be a plc that runs the railway on behalf of the trust. The 'members' of a preserved railway – the subscription payers – usually become members of such a company. Railways might run public appeals, with adverts in the railway press, or there might be a share issue, in which case no dividends are ever paid but the shareholders might get free tickets, perhaps a complimentary meal in a dining car. Their true reward is the perpetuation of a railway, and the trick for the lines will be to find a new generation of people who require nothing more in return for their help.

Meanwhile the original 'Generation Steam' is dying

out, which is an opportunity for the preserved railways, albeit one requiring a tasteful approach. The more dynamic railways issue 'legacy packs' or leaflets. They are distributed around the railways, and perhaps a few will be dropped off in local solicitors' offices. Steve Oates, Chief Executive of the Heritage Railway Association, told me, 'It's thought the next ten or twenty years could be a strong period for legacies.' Steve is a volunteer on the Isle of Wight Steam Railway, which began advertising its willingness to receive legacies about ten years ago. 'It did make a big difference,' said Steve, 'and what's surprising is the way the money sometimes comes from surprising sources – people you didn't know, who were not connected to the railway.'

3

Narrow Gauge in England

The Gravitas of English Narrow-Gauge Railways

Having got our early preserved railways up and running, we will be departing from a chronological scheme in favour of a more thematic one. It seems logical to move on quickly to narrow-gauge lines, because when most people think of a preserved railway, they think of the Ffestiniog, which is a narrow-gauge line. It is also in Wales, where Britain's narrow-gauge lines are concentrated, the landscape requiring trains nimble enough to deal with severe gradients and curves.

England did not have the same concentration of narrow-gauge lines, and nor, in spite of the Highlands, did Scotland, which is surprising until you consider that there wasn't the slate in the Scottish hills that bred the narrow-gauge lines of Wales. The only public narrow-gauge line in Scotland was the Campbeltown & Macrihanish, which carried coal and tourists on the Mull of Kintyre. It was the remotest railway in Britain, and it closed in 1932, soon after the colliery closed.

England has fewer narrow-gauge railways than many European countries, partly because so many standard-gauge lines had been built so quickly – and we ought to say a word about these terms. When we enter the world of the narrow-gauge railway, we are departing from the norm, which is – and has been since the Gauge Act of 1846 – 4 foot 8½ inches, an extrapolation from the width between the wheels of a typical horse-drawn wagon. (It is said that the ruts worn by wagons in the stone paving at Housesteads Fort on Hadrian's Wall are 4 foot 8 apart.) The 1846 Act did not absolutely ban other gauges, and departures from this norm have created wider tracks (the broad gauge) and narrower ones, mainly the latter. Isambard Kingdom Brunel favoured a broad gauge of 7 foot (strictly speaking, 7 foot and ¼ inch) for the Great Western Railway because it gave a smoother ride at speed. But in 1892 the company lost its nerve. In a 'Have-it-your-way-then' hissy fit, the GWR re-gauged its main line from Paddington to Penzance in one May weekend. There are no broad-gauge preserved railways in Britain, although there is a short section of broad-gauge track at the Didcot Railway Centre, where the Great Western is commemorated: a replica of Daniel Gooch's broad-gauge engine, *Fire Fly*, runs there when fully fit.

In the case of narrow-gauge lines, the reason has usually been economy. A narrow-gauge line is anything less than the standard gauge, but there are many narrow gauges, and the category includes 'miniature' railways. The boundary is blurred, but a miniature-gauge railway is usually taken to be less than 15 inches.

Narrow-gauge lines could be built more cheaply because their track bed was narrower, their sleepers were shorter, their rails of lighter section. So narrow-gauge lines are like subsistence hill farming: operating in rough, hilly, economically marginal countryside. They look whimsical, even toy-like, and their rolling stock appears eccentric, often having been built specifically for that line, but they were usually created for the un-whimsical business of freight haulage. If they ran for short distances within an industrial complex they had a good chance of survival, but if they carried their freight over any distance they were vulnerable to road competition.

They tried to compensate by pitching to the tourist market, but the carriage of passengers was no guarantee of survival in the 1920s and 1930s when buses began to probe into narrow-gauge territory, hence the demise of the Leek & Manifold in Staffordshire, which carried milk and tourists, and which closed in 1934, despite the parent North Staffordshire Railway devoting eight of its twenty-three official postcard sets to it. In 1930, it was the subject of a patronising newsreel film (available on YouTube), *A Quaint Little Railway*, the scenarist refining the insult in the first caption to 'Queer Little Railway'. It was said that the Leek & Manifold 'began nowhere and ended up in the same place'. That place was a picturesque valley, but there were not enough tourists to counteract the closure of Express Dairies' creamery at Ecton, and if there is any ghostly echo of the old railway it will be in the form of a nocturnal clanging of milk churns. Other narrow-gauge casualties of the early 1930s include the above-mentioned Lynton & Barnstaple and the Southwold Railway, both of which are remembered today in entirely lyrical terms, and we will be considering the Southwold Railway later on.

The narrow-gauge lines, then, are small not for the reason a lapdog is small, but for the reason a terrier is small. The Ravenglass & Eskdale had started life in 1875 as a 3-foot-gauge line for the conveyance of iron ore over the 8 miles from mines at Boot near Scafell in Cumbria to the port of Ravenglass, and it survived in this way until 1912, when the mines closed. In 1915 Wenman Joseph Bassett-Lowke, best known as a maker of toy trains, converted the line to the 15-inch gauge, and the Ravenglass & Eskdale in this form has been described as the world's first pleasure railway. But even after conversion – when it gained the potentially cute nickname 'L'il Ratty' – it remained tough, in that as well as passengers, it conveyed granite from Beckfoot quarry at Murthwaite. The quarrying ceased in 1953, and the line was saved by the Ravenglass & Eskdale Railway Preservation Society, which bought it from the Keswick Granite Company in 1960.

Today, the R&E is a well-managed tourist attraction. 'Not for the first time in these narrow-gauge travels', Chris Arnot writes in *Small Island by Little Train: A Narrow-Gauge Adventure*, 'I found myself marvelling at how the strain, stress, turbulence and misery of our mining past had somehow been enfolded back into the landscape, leaving hardly a scar behind.'

In the case of the R&E, the landscape is that of the Lake District, which was one of the first railway battlegrounds. William Wordsworth, tenant of Dove Cottage at Grasmere, wrote letters and poems objecting to the arrival in the Lakes of the Kendal & Windermere Railway, which proposed extending a finger into the Lakes from Oxenholme to Kendal, and seemed likely to push on to Windermere, which it never did quite reach, stopping short at a station a mile east of the lake. The line, according to Wordsworth, would bring 'the whole of Lancashire, and no small part of Yorkshire, pouring in upon us to meet the men of Durham, and the borderers of Cumberland and Northumberland'. Here is the famous lyric:

Plead for thy peace, thou beautiful romance
Of nature; and, if human hearts be dead,
Speak, passing winds; ye torrents with your strong
And constant voice, protest against the wrong.

The objection of Wordsworth – who also objected to the introduction of the Penny Post on the grounds that bad poets would start sending him their works unsolicited – was a bit rich, since no one had done more to commercialise the Lakes than he himself, through his publication in 1810 of the bestselling *Guide to the Lakes.*

A modern-day writer about what is now the Lake District is Alfred Wainwright. In 1978 he wrote, for publication by the Ravenglass & Eskadale Railway, *Walks from Ratty*. It begins in praise of Eskdale: 'one of the loveliest of Lakeland's valleys ... unspoilt by commercial and industrial activity ... this perfect Arcadia within the hills'. The railway is co-opted into this Arcadia:

> Eskdale's miles are long and public transport services scanty, so that for many walkers the valley was virtually beyond reach until, in 1960, the inspiration and sacrifice of a group of enthusiasts led to the reopening of the old narrow-gauge railway constructed in 1875 to link the iron mines in the valley with the line at Ravenglass.

The Ravenglass & Eskdale is important in preservation history. As Jonathan Brown writes in *The Railway Preservation Revolution,*

> Railways on the miniature gauge did not have such a

> pool of old locomotives [as standard gauge lines] and had to build new. The lower costs of locomotives on this scale have helped. The Ravenglass & Eskdale built the first of the preservation era, the 2–8–2 locomotive *River Mite* of 1966. *Northern Rock* followed later ... Locomotives for the Bure Valley Railway have also included many of new construction.

The Bure Valley Railway – opened in 1990 – is a 15-inch line built along the trackbed of a standard-gauge railway. It occupies a 9-mile stretch (Wroxham to Aylsham) of what started life as a branch of the East Norfolk Railway. The branch, which mainly carried coal, closed in 1982 and the track was lifted in 1984. It runs through countryside that is, for Norfolk, unusually hilly or at least undulating, and it can boast the Aylsham Bypass Tunnel, the longest operational railway tunnel in Norfolk. Then again, there were only ever two (compared to thirty-seven in Yorkshire), the other one being the Cromer Tunnel, which is disused. But where there are few tunnels, there are likely to be many bridges, and the Bure Valley is traversed by seventeen.

There are many impressive facts about the Bure Valley. It has an unusually high number of paid staff alongside its team of volunteers: seventeen. The well-stocked bookshop at the Wroxham end is open every day of the year except Christmas Day and Boxing Day. The same goes for the model railway and toy shop at the Aylsham end.

This railway was not financially viable until 2001, when it was taken in hand very firmly indeed by a married couple, Andrew Barnes and Susan Munday. He was a rail enthusiast who had been a City banker, she an accountant. They run the line as a not-for-profit company and have invested heavily in it. The carriages are very solid, and the one I rode in seemed to be heated even on a sunny day of riding by vast golden cornfields. (There will never be any open-sided or open-topped carriages on the line because of the tunnel.) Aylsham station, which looks like a giant bungalow, is bigger than most standard-gauge stations on the national network. There's a locomotive depot, an engine shed and a very clean workshop, in which a locomotive was mounted on a work table like a patient in an operating theatre. The staff wear blue shirts with the BVR insignia; the same insignia appears on the sunshades outside the large restaurant at the station. The size of the car parks at Aylsham and Wroxham shows the ambition – and success – of the Bure Valley Railway.

As given on its website, the history of another 15-inch railway, the Kirklees Light Railway in West Yorkshire, suggests a sudden loss of gravitas. The account opens with a description of the tortuous railway politics that led to the construction, by the Lancashire & Yorkshire Railway, of a branch off the Penistone Line between Huddersfield and Sheffield. The branch was primarily for the transportation of coal. It closed in 1983, and the track was lifted three years later. Now comes the change from a major to a minor key: 'Around this time, Brian and Doreen Taylor had established a miniature railway at Shibden Park in Halifax. This had become a great success and continues to please many visitors to the park today. Brian however wanted to get his teeth into something bigger, and began a search for somewhere to build a 15-inch gauge railway.' But then we come full circle, back to weightiness, given that all the engines on the reopened line were built by Brian Taylor; and the line includes the Shelley Woodhouse tunnel – the longest on any 15-inch gauge railway.

Construction of the Kirklees Light Railway was authorised in 1991 by a Light Railway Order, one of the last given under the Light Railways Act of 1896 – which seems a fitting moment to say something about that legislation, which permitted a railway under laxer, or less stringent, specifications than a 'heavier' railway. The line might be exempted from fencing or station platforms; level-crossing gates might not be required. In return, a speed restriction was imposed. A light railway built under the Act did not need an Act of Parliament, only a public enquiry conducted by the Light Railway Commissioners. In spite of the Act, there is no definition of 'light railway' any more than there is of 'light music'. A Light Railway under the terms of the Act was not necessarily narrow-gauge, although the Act did underpin many of Britain's narrow-gauge lines, but not every line called a 'light railway' necessarily conforms to the terms of the Act. The original context was glum: the Act was designed to help farmers get their goods to market at a time of agricultural depression. Relatively few lines were built under the Act, and most were short-lived. Something accidentally joyous and charming did come out of the Act in the form of the eccentric Colonel Stephens' lines, which will be considered later.

The Act, or its successor, the Transport and Works Act 1992, also governed most of the preserved railway re-openings.

The Romney, Hythe & Dymchurch Railway

The Captains

Broadly (as it were) speaking, a narrow-gauge line is anything less than the standard gauge, but there are many narrow gauges, and, as we have said, the category includes 'miniature' railways. A miniature-gauge railway is usually taken to be 15 inches or less, and its rolling stock usually comprises small versions of full-sized prototypes. We will stick with the miniature gauge, as we enter the patrician world of inter-war estate railways.

These are evoked in black-and-white images of full-grown men, usually in tweed suits and ties and smoking pipes, sitting on trailers attached to locomotives about the size of dachshunds, with commensurately small carriages or wagons behind, usually adapted for the carrying of people, in which case children might be aboard, but these are usually in the background, out of focus. All attention is on the chap on the engine, who never shows any sign of 'seeing the funny side', perhaps because his locomotive might have cost – in the case of, say a 7¼-inch-gauge 4–4–0 – anything from £12,000 to £50,000 in today's money.

The engine had very likely been supplied by the firm of Bassett-Lowke, founded by the above-mentioned Wenman Joseph Bassett-Lowke, who was the son of a boilermaker, but had an aristocratic style, in accordance with his name. He was, like that other aesthete of the railways, Frank Pick (deputy head of the London Underground in the inter-war years), a member of the Design and Industries Association, which sought to bring Arts and Crafts values into the machine age, and he was a mentor to Charles Rennie Mackintosh. Bassett-Lowke had begun by importing steam-driven or clockwork 'carpet' locos (they could be run over a track laid on carpet) from Nuremberg, where model engineering had grown out of clock-making. After the First World War, the imprimatur 'British Made' was sought by patriotic purchasers of model railways, and Bassett-Lowke manufactured his own models in association with his business partner Henry Greenly, a graduate of the drawing office of the Metropolitan Railway. Bassett-Lowke had a shop on Holborn, near the junction with Southampton Row, the window advertising, in elegant Johnston sans serif, 'Model Railways, Ships, Architecture'.* Note 'model' rather than 'toy' railways.

A maker of *toy* railways was Frank Hornby, who believed that 'Play is the work of childhood' and who, inspired by his love of watching the cranes at Liverpool Docks, had invented Meccano in 1901. His model railway business was an outgrowth of Meccano, and in 1938 he came up with the world-beating electric Hornby Dublo sets, in the Double-O gauge. These must have been owned by many of the boys who would go on to preserve full-sized railways, which might be considered to be the ultimate model railways. But Bassett-Lowke had a more direct involvement with preserved lines, through his commissions from the gentry.

Some of his customers, admittedly, were merely 'Mr'. There was Mr Drysdale Kilburn, who ran an orchestra which, according to *Miniature Railway Magazine*, issue 40 (Spring 2018), 'was responsible for the first broadcast of Elgar's music in 1923'. He built a 7¼-inch-gauge line

*It is now, of course, a McDonalds.

at his home, Derby House in Hendon, with a Bassett-Lowke replica of LNWR engine *George the Fifth* as motive power. But most of Bassett-Lowke's clients were titled in some way, even if only 'Captains'. There was Captain J. A. Holder, who ran a 10¼-inch- gauge line around the garden of his home at Broome, near Stourbridge. Another client was Captain Charles Frederick Warde-Jones, who ran a 10¼-inch-gauge railway in the grounds of his estate at Harness Grove, near Worksop. There were, according to *Miniature Railway* magazine, 'bridges of miniature brickwork and embankments, with a double-track main station complete with a refreshment room selling chocolate, cigarettes and drinks'.

Above all, there was Captain John (Jack) Edwards Presgrave Howey, the Old Etonian heir to a fortune made in Australian real estate, who – back when he was still Mr Howey in 1909 – had built a 9½-inch-gauge line in the grounds of his home, Great Staughton Manor, near St Neots. After the war, and with the assistance of Bassett-Lowke, he converted the line to a gauge with an august pedigree: the 15-inch.

Sir Arthur Heywood (1849–1916), a skilled amateur engineer (in spite of being a graduate of Eton and Cambridge, and a baronet), asserted that 15 inches was the 'minimum gauge', meaning the narrowest track that could carry people in some degree of comfort. He built a line of that gauge in the grounds of his estate, Duffield Bank, near Derby. He had hoped his ideas might have some application in military or agricultural use, but his only client was the Duke of Westminster, for whom Heywood built the 4½-mile, 15-inch-gauge Eaton Hall Railway in Cheshire. Heywood built a sleeping car and a dining car in the 15-inch gauge, but estate railways are not necessarily just for dilettantes: they do serve the estate, and the Eaton Hall Railway connected the house to the GWR at Balderton (albeit not directly, because of the clash of gauges).

The espousal of 15 inches was perpetuated by Bassett-Lowke, who bought up the Duffield Bank equipment after Heywood died, and converted the derelict Ravenglass & Eskdale Railway in Cumbria to 15-inch gauge, as we have seen. A year later, in 1916, he also converted the Fairbourne Railway, which runs for a couple of miles along the beach from Fairbourne to Barmouth Ferry railway station. This anticipates our trip to Wales, but a word here about the rather slippery Fairbourne. It was always for tourists, and it started life as a 2-foot-gauge horse-drawn tramway. It prospered for a while after its conversion by Heywood, but it closed in 1940. It was revived by a consortium of Midlands businessmen in 1946 – early for a preserved railway; but *is* it a preserved railway? In 1984, it was acquired by the Ellerton family and converted to 12¼ inches. In 1995 it was acquired by Professor Tony Atkinson and Dr Roger Melton, an impressive-sounding pair. In 2008, the line was transferred to a charity. There is – and has been since the 1970s – a supporters' society, but the line has essentially been a commercial operation with a paid staff, and the Fairbourne Railway is not in the Heritage Railway Association.

Captain Howey was not satisfied with running an estate railway, even if it was on the 15-inch gauge; even if it did feature a scale model of the Forth Bridge, and even if he had commissioned from Bassett-Lowke a really enormous small engine to run over its metals: this was a loco named *Gigantic*, and it was the second-ever engine to be built in Britain with the Pacific wheel formation of 4–6–2. Howey wanted a 15-inch line that was at large in the world, a main line in miniature. His co-conspirator was Count Louis Zborowski, a descendent of the Polish aristocracy on his father's side and the Astors on his mother's. Like Howey, he had a Bassett-Lowke 15-inch-gauge model railway on his estate at Higham Park near Canterbury; the motive power was an engine now called *Count Louis* that is today at the Evesham Vale Railway, which runs through an orchard in Worcestershire. Also like Howey, Zborowski was a racing driver, even though his father had been killed racing in 1903. Zborowski designed and owned two racing cars nicknamed *Chitty-Bang-Bang*, the name inspiring Ian Fleming (along with his admiration for Zborowski) to write his three children's books featuring *Chitty-Chitty-Bang-Bang* (he added an extra *Chitty*), which are dedicated to Zborowski. Howey and Zborowski met at Brooklands in Surrey, which was a racing circuit with a *social* circuit. 'Races took place most weekends during the season,' writes Charles Jennings in *The Fast Set*, 'but, like a golf course, the place was open all throughout the week for whatever a motoring gentleman with time on his hands might feel able to enjoy.' In 1924, Count Zborowski was killed when his car hit a tree at the Monza Grand Prix, so it was left to Captain Howey to build a line on which their 15-inch locos could have a proper work-out.

The route he chose was on Romney Marsh in Kent, between Hythe and New Romney, the termini of two Southern Railway lines that the Southern had no plans to connect directly. There was a business case for connecting the two places, but not much of one, otherwise the Southern *would* have built a connecting line. The main purpose of the projected line was to fulfil the dream of Captain Howey. Howey bought up the land, and the first public train ran the 8 miles from Hythe to New Romney on 16 July 1927. The line was double-tracked, partly so that the Captain, driving one of the locos, could race a friend driving another. The impatient and determined Captain opened the 5-mile extension to Dungeness two years later.

Howey not only built the railway, he also imperiously managed it. In looks, Howey was wolfish, like the actor Ralph Fiennes, and he is never smiling in photographs. *Miniature Railway* magazine suggested he was camera-shy, but he also had a flamboyant side. He enjoyed theatre-going and would command the staff of his railway to go off and see a show he had enjoyed, while he held the fort. Captain Howey died, in his bed, in 1963. The subsequent fortunes of the line are summarised by Denis Dunstone in his book *For the Love of Trains: The Story of British Tram and Railway Preservation,* and I quote this passage with reference to the controversy about whether the RHD really is a preserved line:

> It is questionable whether it is a preserved railway at all for, apart from during World War II, it has been in continuous operation since it was first opened in 1927. However, after the death of its founder and owner Captain Howey in 1963 it very nearly collapsed, and was only saved by a consortium formed by Bill McAlpine in 1973. For this reason it can be regarded as a preserved railway and it has certainly regarded itself as such, being a fully paid-up member of the

Heritage Railway Association (HRA), while its managing director for many years, John Snell, served as Deputy Chairman of the HRA.

But, as we saw in Part One of this book, not everyone involved with the railway does regard it as preserved, and if it *is* preserved, it is one of the few in that category to be run along commercial lines.

Down the Rabbit Hole

There is something generally outrageous about the line Howey created, starting with its name, which is very long for such a short (14-mile) line;* and it is not correctly ordered, since the railway proceeds southwards along the coast from Hythe to Dymchurch, New Romney, Dungeness. So it should be the Hythe, Dymchurch & Romney Railway. Perhaps the total exclusion of Dungeness, a place name that – with its echo of 'dungeon' – would be baleful enough even were there not a nuclear power station on its desolate grey beach, is for PR reasons. I do not say 'bleak' because there is a guest house on that beach with *objets d'art* in the back garden, and one of those is an old-fashioned peg board of the kind that used to announce dishes of the day in transport cafés, and the wonky letters insist: 'Dungeness is not bleak.'

My journey began at Hythe, the main terminus, where the RHD has four platform roads – an impressive railway set-up, but with a slight suggestion of a garden centre, with lineside plants in pots, and ceramic animals for decoration. The waiting engine was based on a Canadian

*Then again, it is the longest 15-inch-gauge line in the world.

Pacific and built by the Yorkshire Engine Company in 1931. It had a cowcatcher on the front and was called *Dr Syn*, after a character in a series of novels by Russell Thorndike beginning with *Dr Syn: A Tale of the Romney Marsh*. The Reverend Doctor Christopher Syn is a brilliant eighteenth-century scholar and swordsman who, embittered by the infidelity of his fiancée (or something like that), turns outlaw and smuggler. His adventures were filmed three times, most famously and ridiculously by Disney as *The Scarecrow of Romney Marsh*.

The driver of *Dr Syn* was polishing the boiler with Johnson's Pledge. In an essay about the RHD in *Railway Roundabout: The Book of the TV Programme*, Bill Hart wrote, 'The Romney driver's life can be summed up in three words – cleaning, cleaning, cleaning. Mechanical soundness and reliability are as near perfect in these engines as constant vigilance can ensure.' He argued that this attention to cosmetics stemmed from the lack of the usual narrow-gauge setting: the conspicuous lack of 'mountain passes, precipitous ledges, narrow gorges and the like ... There is no magnificent frame to the picture, and so it must be the picture itself that provides the appeal.'

Dr Syn, like all the Romney steam engines (there are also some diesels), is about 4½ feet tall to its chimney top, which is bigger than it might have been. Whereas the track, at 15-inch gauge, is about a quarter of full size, the engines are about one-third full size. The fire hole door was the size of a cat flap; the shovel was the size of a child's in a beach bucket-and-spade set. The driver sits down in the cab, but he is not actually a driver: he is a 'motorman', who works alone with no fireman.

Some of the carriages had open sides, and some were

enclosed but – as it turned out – draughty. Sitting down, I felt like Will Ferrell in *Elf*, too big for my surroundings, and again the question came up: how real is this railway? The RHD carries 125,000 passengers a year and turns over £2m; it has about forty paid staff – a high number – so there's a touch of the Fairbournes about it. Tourism has always been its main prop, especially in the inter-war years when a number of holiday camps sprang up along the line, but it *is* used for purposes other than leisure. Locals can buy shares in the railway, thus acquiring the right to free travel, which they use to go shopping in Hythe. Between 1977 and 2015 it served a school – John Southland's Comprehensive at New Romney – but, in a *Titfield*-like scenario, the railway's school run was undercut by that of a bus company. It must have been odd to be travelling on the train on the way to doing your GCSEs: the combination of the apparently frivolous and the serious. The RHD doesn't run every day of the year, but for more days in winter than most preserved lines.

The train ambled away from Hythe in a grey drizzle, passing the backs of modest houses. The householders seemed in survivalist mode, with caravans, upended boats and gas canisters in the gardens, together with outhouses from which flags of St George defiantly fluttered. The houses gave way to marshland, with the English Channel seeming always just out of view to the left, but perhaps it is the proximity to France that explains the England flags, and creates a sense of vulnerability. In the war, the RHD was taken over by the military, who operated a fully armoured little train as part of coastal anti-aircraft defence, and how *Dad's Army* writers missed that plotline I do not know.

We came into New Romney, another impressively

rangy set-up, and the location of the RHD's workshop. Above the station buffet was a room accommodating a sprawling HO-gauge model railway, and other models of different scales, some saying 'Not to scale'. The effect was disorientating. I kept reading placards like, 'The scale of 16 mm to the foot loosely represents 2-foot narrow gauge using "O" gauge track.' A green tank engine in a glass case was described as a model in 'unusual' 10⅛th gauge, of an engine that 'never existed in real life'. Here was a railway world out of the mind of Lewis Carroll, and it seemed fitting that humourists have played roles in the RHD story. The model railway room featured a photograph of Michael Bentine, who was too eccentric even for the Goons (he dropped out of the ensemble early on), but who was the honorary vice-president of the RHD in the early 1990s. He is pictured on a visit to the line with his erstwhile colleagues Harry Secombe and Spike Milligan. There is also a photograph of Laurel and Hardy wearing silly hats – Stan's too big, Ollie's too small – while pretending to drive an engine and making even Captain Howey smile. They had been brought in to perform the 'formal reopening' of the line after the wartime closure, and I like the notion of Laurel and Hardy doing anything formal.

Perspectival disorder continued as I stepped back onto the station. Platform benches appeared to be normal-sized, but what about a sign reading 'Danger: Low Bridge'? Surely that ought to read 'Danger: Low Bridge (even by the standards of this railway)'? A poster advertising a walk between Hythe and Sandling Station (which is on the national network) was headed 'Big Train, Little Train'. I stepped out of the station briefly to see the

'Captain Howey Hotel', but it was just an ordinary semi-detached house.

Another poster depicted an RHD train above the slogan, 'Smallest public railway in the world', and this triggered the slightly testy discussion with the RHD volunteer I mentioned in my introduction. The RHD certainly is a public railway in that it was the subject, on opening, of a Light Railway Order under the Light Railways Act of 1896. But that word 'smallest' ...

I asked the volunteer, 'What about the Wells & Walsingham? That's only four miles long, so it's shorter than the RHD, but it's also *smaller* –10¼-inch-gauge – and it too was authorised by a Light Railway Order in 1982. So surely *that's* the smallest public railway in the world?'

My interlocutor deftly threw the blame for any confusion on the European Union: 'It's the sort of thing EU bureaucrats could argue about for years, and I'm sure they'd love to.' (We will be visiting the Wells & Walsingham very soon, by the way.)

The train I caught from New Romney was hauled by an engine called *Northern Chief*. (Miniature-gauge engines do tend to have these grandiose names, as if they suffer from Little Man Syndrome.) There were two men squeezed into the cab, and I wondered if this was a contradiction of the RHD rule of one-person operation, but the driver, indicating his mate, said, 'He's just there to look pretty.' *Northern Chief* is one of five RHD engines dating from the mid-1920s that are based on Gresley's A3 Pacifics for the LNER. Apple-green *Typhoon* is really the *Flying Scotsman* in miniature, but it's not called that, out of respect for the bigger engine. In 1927, newly built *Typhoon* was taken to King's Cross Top Shed along with a length of demonstration track to be photographed alongside the real thing, which had been built four years previously. A few years ago, when *FS* was visiting the Bluebell, *Typhoon* was taken down to that railway – along with its display track – for the photograph to be re-created.

We rolled past some more back gardens, then we were into the mesmerising wilderness of Dungeness, which the RHD traverses in a great, apparently crude loop of the kind I used to make on the carpet when I had a model railway. (That was when I was a boy, by the way.) But is 'wilderness' the term for this headland of shingle? It is officially a desert, but also a site of Special Scientific Interest with a huge variety of flora and fauna, most of it managing to blend in with the greyness. It seems that colour is a luxury here, the whole place being monochrome, the tone set by the Edwardian lighthouse, which is black and white like the pole of a Belisha beacon. At Dungeness, it is as if the catastrophe the householders along the line had been preparing for had actually occurred. The few scattered dwellings are of black wood, as though charred in some great conflagration that wiped out most of their neighbours. They stand amid upturned boats, abandoned shipping containers and other bits of industrial flotsam. Some of the houses are built around a core structure of standard-gauge railway wagons, testament to the fact that Dungeness was once the terminus of a big railway branch (closed to passengers in 1937) connecting it with Appledore. The station for that line – now demolished – stood across the road from the lighthouse. Part of the branch is used to take nuclear waste away from the Dungeness nuclear power station, which – uningratiatingly massive and blank – dominates the scene, suggesting itself for the role of guilty party in this apparently post-apocalyptic scene.

In case any of this is coming over negatively, I ought to say that Dungeness, like the trains that serve the place, is both fantastical and gloweringly beautiful. As the enormous sky darkened, it seemed to herald the brewing of a conspiracy between the grey sea, the wind and the nuclear power station, and I made very sure I was back at the station in time for the last train back to Hythe.

Three Narrow-Gauge Seaside Lines in England

The Wells & Walsingham Light Railway

The English narrow-gauge lines that have survived have often done so because they are in a tourist catchment area – like the seaside.

I turned up at the Wells & Walsingham Light Railway in heavy rain one day in late March. I parked in a muddy puddle next to the small, brick-built station, which had a defiant church fête air, with bunting being lashed about by the wind, and a multi-coloured banner reading 'WE'RE ALL MAD HERE'.

As mentioned, the Wells & Walsingham Light Railway (WWLR) is four miles long. It occupies the northern part of what was once the standard-gauge Wymondham–Wells line, built by the Norfolk Railway (and a couple of others) and later absorbed into the Great Eastern, then the London & North Eastern. Then came BR, and the Beeching axe. I think of the Wymondham-to-Wells line in its modern incarnation as being like a space rocket. At the base, or the southern end, are the thrusters, and these are provided by the big diesels of the preserved standard-gauge Mid-Norfolk Railway, which operates between Wymondham and Dereham. At the top, or northern, end is the smaller kit: the lunar module, which we might think of as being any one of the little engines of the 10¼-inch-gauge WWLR as they wobble towards the stratosphere (or the North Sea). This sense of the WWLR fulfilling a lonely mission was inculcated in me by the long drive past vast, low, sodden fields that had taken me from Norwich to Wells, and my awareness of the fact that, for all the northbound probing of the preserved Mid-Norfolk, it has no plans to go as far north as the WWLR's southern terminus of Walsingham. So there will never be any interchange between the standard and narrow gauges, and the WWLR will remain out on a limb.

The WWLR station called Wells-on-Sea is not quite in Wells itself, but in a patch of scrubby countryside a little way south of it. The station is separated from the town by the A149, and the incessant, baleful swishing of wet car tyres was a painful reminder of the fact. 'If we could move that caravan, and all that other junk', a volunteer platform guard told me, indicating the road, 'that's where the crossing-keeper's cottage was.' The building that housed the original Wells station of the Wymondham–to–Wells line was also called Wells-on-Sea, and it is now occupied by a seller of second-hand books. It is located fairly centrally in Wells, but is not on the seafront, which is why, shortly before its closure, BR had a fit of conscience and renamed it Wells-*next*-the-Sea (as opposed to 'on'), which is the official name of the town, the sea being rather elusive off this coast, especially at low tide.

The volunteer guard would love the WWLR to

somehow cross the road and go into the town once again, ideally proceeding past the old station and towards the sea. 'Then we could take people to the beach. Imagine that!'

I imagined it: the WWLR could be a sort of Park & Ride for Wells, if they filled in the craters in the car park. The WWLR could also provide a means of direct access to Wells for the people of Walsingham. For most of the year, that means not very many people, the population being about 800, but it is a village of shrines and is frequently full of pilgrims, some of whom do end up relaxing on the beach at Wells, after a hard day's praying.

The begetter of the WWLR – its own Captain Howey – was Lieutenant-Commander Roy Francis (1922–2015), who saw very active naval service during the war, being torpedoed four times. On leaving the Navy in 1956, he built boats on the Norfolk Broads, and, according to his obituarist in the *Daily Telegraph*, 'He built a railway in his garden for his own amusement, and with his son, a medical student, and his daughter, a ballerina, he took a portable track to fêtes and fairs to help pay the bills.' In the early 1970s, he was employed by Norfolk County Council to ease traffic congestion by the construction of a 10¼-inch-gauge line along the flood bank between Wells Harbour and the caravan site (and beach) at Pinewoods. This job made him acquainted with the trackbed of the old Wymondham–Wells line. With leases from local landowners and the County Council, and his Light Railway Order secured, Lt-Cdr Francis began construction of the WWLR, which opened in 1982, with motive power in the form of a steam engine appropriately named *Pilgrim*.

The line was an immediate success with both trippers and locals, to the extent that a more powerful engine was needed to haul longer trains. So Francis commissioned the building of a Garratt locomotive. We will say more about Garratts when we come to the Welsh Highland Railway, but for now let's say that they have two boilers. Its name was *Norfolk Hero*, and this was the engine on duty when I turned up at Wells-on-Sea Station.

The principal structures of the station are an old standard-gauge carriage, which bears a sign reading 'This carriage dates from 1900 and will form a café when restored,' and a (full-sized) two-storey signal box, the upper part of which is used to control signals, while the lower part is the café and shop. This signal box used to be on the Colchester–Norwich line at Swainsthorpe and became surplus to requirements when that line was electrified. It was lifted onto a lorry intact and driven to the WWLR. According to the official history of the line, 'Later that day [it does not say which day] Norfolk people were treated to the sight of a signal box driving past their front doors.' But I think the sight would have been even more surreal if the box had been carried by rail, because then it would have seemed to the casual observer as if the railway itself was moving. The other travails of construction and operation are described in the official history, including the construction of a new engine shed, the old one having begun to lean to the right.

In 1993, some new carriages were built, with windows closable in the Victorian manner. You lift them by hauling on a leather strap resembling a barber's strop, which you then hook into place by aligning a hole with a brass knob. According to the official history, 'These have proved very popular in all weathers and ... look very elegant indeed alongside *Norfolk Hero*.' I certainly agree with the last part, and would be reluctant to question the first, but I found the carriage draughty and my feet were cold – possibly my own fault, since my boots had got waterlogged when I'd stepped into the car-park crater.

Even so, the trip was enjoyable. As always on a narrow-gauge line, 20 miles an hour feels more like 60, and the countryside was unspoilt, with a melancholic, low-key allure. Immediately out of Wells and to the right, we ran through a copse that contained a wooden house like something out of a fairy story, and there was a big bear of a man sitting outside it and chopping wood; he waved and beamed at us as we went past.

The WWLR station at Walsingham is really just a halt; it is not the old station of standard- gauge days, because that is now St Seraphim's Chapel, 'a centre for Orthodox Iconography', which also houses some railway memorabilia. In any event, it was closed.

When I returned to the train, the driver – in spite of the rain – was laughing on the platform with some members of the public. He seemed so happy in his volunteering that I was quite jealous, as I often am when encountering preserved railway drivers. I asked how long it would take to learn to drive *Norfolk Hero*.

'Oh, a couple of seasons. Depends very much on aptitude.'

On the way back, I sat in one of the open carriages, reasoning that it couldn't be much colder than one of the enclosed ones, and might be slightly warmer, given that thick woollen blankets are offered to those in the *plein air* carriages. I did benefit from the sauna-like effect of the steam clouds from *Norfolk Hero* that periodically enveloped me.

After my ride on the WWLR, I walked into Wells-next-the-Sea, calling first at the station of the big railway that is now a bookshop. It was a rambling place, with several rooms, in which there were handbells to summon the proprietor from his private quarters, which were perhaps a bit warmer than the downstairs ones. He had a good selection of green Penguin thrillers.

I walked on into the town, and as far as the harbour, where the tide was out – a very long way out. Rain fell on puddles of seawater and marooned boats. I carried on to the terminus of the above-mentioned Harbour Railway, which took the form of a concrete platform and a garden bench. The train was approaching: three people stepped off it. There would be one last outbound trip of the day to Pinewoods although, as the driver explained while turning the engine around on a turntable with a circumference of a dustbin lid, I would have to walk the mile back. The engine was blue and called *Howard*. It was a diesel, made to look like a generic steam engine. At the end of the line, the driver drove *Howard* into an engine shed that looked like the garage of an ordinary suburban house.

You could see why, having built this, Lt-Cdr Francis would want to move on to the challenge of a bigger small-gauge railway with all the enjoyable clutter and untidiness associated with live steam, but the Wells Harbour Line has its own claim to fame: when it opened in 1976, it was the first 10¼-inch-gauge railway to run a scheduled service.

The North Bay Railway at Scarborough

The subtitle of the souvenir guidebook of the North Bay Railway at Scarborough is *A railway for pleasure since*

1931. The railway begins next to a boating lake, and runs amid innocent, salubrious attractions – principally flower beds, but also an open-air theatre and a golf course – for seven-eighths of a mile (a charming distance) along Scarborough's North Cliff to the suburb of Scalby Mills. There is a YouTube film of the line, and some commentators on the film condemn the railway for being a swizz – using engines that look like steam engines but aren't. But the more thoughtful and better-informed people point out that the engines are (a) beautiful and (b) historical, in that they are not only early diesels, but also early examples of 'steam-outline diesels': diesels designed to look like steam engines in almost every external particular, and if this is a deception it is an honourable one, like *trompe l'oeil* painting.

The first batch of engines – and the core of the present fleet – was built by Hudswell, Clarke of Leeds, who were suffering the effects of the Depression at the time and needed the work. The engines were built to resemble those local heroes of the London & North Eastern Railway, the Gresley Pacifics. Hudswells themselves recommended steam-outline diesels as more fitting to the gentility of the locale: no coal would have to be brought into the gardens; no smoke would sully the flowers ... and the engines featured a form of hydraulic transmission provided by a Vickers-Coates torque convertor. According to the Hudswell, Clarke prospectus: 'The entire absence of uneven take-up and "snatch" or "grab" makes this torque convertor unique. The driver has absolutely no reason to gently open the throttle. He can open it as quickly as possible, or slowly; no damage is possible, and the driver has a carefree mind with the convertor.'

But things have not always gone so smoothly on the line: in 1932, the second operating season, a head-on collision killed the driver of one of the engines, Herbert Carr, aged twenty-five. This incident is not mentioned in the souvenir guidebook.

I have ridden many times on the North Bay Railway. When I was a boy, a trip along the line was usually a reward after a family swim in the waters of the open-air North Bay Bathing Pool, which were freezing even on the hottest day.

The Lincolnshire Coast Light Railway

We should not leave the seaside without mention of another narrow-gauge holiday line: the Lincolnshire Coast Light Railway. Like the North Bay Railway this is a member of the Heritage Railway Association. It is not precisely a 'preserved' line, but it is important in the history of preservation. It crops up in a short, statistically minded survey, *Railway Preservation in Britain 1950–84*, published in 1985. The work, by Eric Tonks, is very dry, but here is Tonks in effusive mode:

> The success of these two lines [the Talyllyn and the Ffestiniog] could not but strike a responsive chord in enthusiasts everywhere, eager to have a go at running their own narrow-gauge railway; and the very limited supply of these left only one alternative – to build their own. If asked which was the first of the 'new' lines to be laid, it is unlikely that many would guess correctly that it was the Lincolnshire Coast Light Railway ... Later schemes have rather eclipsed the pioneer efforts of this remote little line in terms

of public appeal, but none can rob the LCLR of its priority.

The Lincolnshire Coast Light Railway was not – like the Ravenglass & Eskdale or the Wells & Walsingham – a narrow-gauge line built on the trackbed of a standard gauge line. It was a brand-new railway, built on land leased from Grimsby Rural District Council, and opened in 1960 by rail enthusiasts who'd got their hands on the equipment of the Nocton Potato Estate Railway, part of a network of narrow-gauge lines built to deliver potatoes to standard-gauge railheads in Lincolnshire. The wagons had previously been used to carry men and armaments on the Western Front. The railway has moved several times along the Lincolnshire coast. Today it is at Skegness Water Leisure Park, at Ingoldmells, just north of Skegness. Ingoldmells is a problematic spot, overwhelmed with static caravans, but (if I remember correctly) you can't see them from the pleasant environs of the Water Park.

4

Narrow Gauge in Wales

The Ffestiniog Railway

Porthmadog

I drove to Porthmadog. I blame Dr Beeching for making the place so inaccessible by rail from London. Before his Report, you could go in an elegant diagonal to North Wales (Euston, Bangor, Caernarvon). Now, the least worst option would be to go to Shrewsbury, then hard left to the coast before heading north from there: a series of right angles.

I arrived in Porthmadog the night before my proposed trip along the length of the Ffestiniog line, which would involve taking the 10.05 from Harbour Station in Porthmadog to the terminus at Blaenau Ffestiniog. But I slept in the following morning, so it was 10.02 when I pulled into the station car park, in which there were no free spaces. But on hearing that I was an author, the volunteer car park attendant (who, judging by his well-spoken competence, might have been anything from a retired GP to a retired JP) let me park in a space reserved for VIPs. This was the Ffestiniog in typically efficient and accommodating form. I ran onto the platform just as the platform guard was giving the train the 'Right away'. I moved to open a carriage door, but the guard commanded me politely (if loudly) to desist, so the train pulled away without me. I was sorry to miss it, because I'd have to wait an hour and a half for the next one, and the train had been hauled by a Double Fairlie locomotive.

The track gauge of the Ffestiniog (1 foot 11½) is narrower than that of the Talyllyn (2 foot 3), but its track-work and rolling stock are larger and heavier. The Double Fairlies – pioneered by a Victorian engineer of that name – are designed for maximum power on the narrow gauge, and they manage this by being two engines in one. Their appearance is bizarre, like the Pushmi-pullyu of the Dr Dolittle stories. It's as if two tank engines have run bunker-first into one another and fused together: they have two boilers, one firebox, two regulators (levers that admit steam to the cylinders), but only one reverser, so they can't split themselves in two. Their wheels are on bogies – units that swivel to deal with sinuous mountain routes – and all the wheels are powered. The Ffestiniog currently has two Double Fairlies, one of which, *David Lloyd George*, it built itself. They are the talismans of the Ffestiniog. A typical film clip of the line will show a Double Fairlie in operation and, after the viewer has been given time to marvel at its strangeness, only then does the camera pull back to show the magnificence of the setting.

The car park attendant suggested a coffee and a bite to eat – 'It'll be on us' – in Spooner's Café, which is named after the Victorian family that constructed and managed the Ffestiniog and which, with typical Ffestiniog thoroughness, is billed as a café, bar *and* grill, and is 'open early 'til late seven days a week'. When I told him I'd been hoping to see a press officer he immediately paged that department. An interview could take place in five minutes' time, meanwhile could he show me around? Thanking him, I looked for myself.

Harbour Station – adjacent to the harbour – is slate grey because it is built of grey slate. It is the HQ of the Ffestiniog Railway, and the beating heart of Porthmadog. It makes the town (aesthetically not very distinguished) seem important and festive. As you walk the crowded main street of Porthmadog, you imagine that everyone is walking towards or coming from the station, and sometimes the trains are even in the high street, as we will be seeing. The station is the hinge of two preserved 1-foot-11½-inch lines: the Ffestiniog and the Welsh Highland Railway. The Ffestiniog Railway Company operates both, which together constitute North Wales's leading tourist attraction.

The Welsh Highland Railway is 23 miles long, going broadly north from Porthmadog to Caernarvon; the Ffestiniog is 13 miles long, heading broadly east, from Porthmadog to Blaenau Ffestiniog. In high summer, when the most intensive timetable is in force, you can travel there and back on each line in one day, but you would spend eight hours on a train, changing at Porthmadog. You always have to change at Porthmadog, which is good news for Porthmadog.

Clare Britton is the Ffestiniog Railway press officer. Her mother had been a volunteer fireman on the line in the late 1960s, while the family were living in Nottingham: 'We had working holidays here at Porthmadog with the East Midlands Group working party.' Well, Clare and her mother did – her father, a solicitor, would not come: 'He called it "that bloody railway".'* Many supporters

*Many lawyers, and other professionals, did come, however. In his

and volunteers on the Ffestiniog are women. 'Unusually for a preserved railway, we have a good gender balance, and a lot of the Deviationists were women.' (We will be coming to the Deviation.) 'We currently have one female driver on the railway, and one in training.' The Ffestiniog reaches widely for its 1,000 active volunteers and 8,000 members: 'There's a White Rose group in Leeds; a group at Bristol, another at Gloucester.' But it also makes a point of recruiting locals into its 100 paid staff. 'We're advertising for a marketing assistant in the Porthmadog Job Centre right now.'

Clare gave me several documents relating to the history of the Ffestiniog, from which I offer this distillation.

The Festiniog Railway Company was founded by Act of Parliament in 1832, making it the oldest operational railway company in the world, according to the *Guinness Book of Records*. Ffestiniog was spelt with only one F in the Act. In the 1970s, the Welsh spelling, with two Fs, began to be used, and that is the common usage today, but it would take another Act of Parliament to change the official name, and I'm going to use two Fs from now on.

The line opened in 1836. At first, horses hauled wagons uphill from the port of Porthmadog to the slate quarries around Blaenau. The wagons, loaded with slate, ran back to Porthmadog by gravity, under the charge of a brakesman, with the horses riding in special trucks called dandy

---

1975 book about the Ffestiniog, *The Little Wonder*, John Winton characterises the early volunteers as 'in the great majority middle-class ... Schoolmasters and civil servants acted as guards. Surgeons could be seen manhandling sleepers, wearing gloves to protect their scalpel hands.'

cars. The trains took six hours to reach Blaenau, half an hour to come back. Steam engines were introduced in 1862, which is the cue for another fanfare: the Ffestiniog was the first narrow-gauge steam railway. In 1865, it became the first narrow-gauge railway to carry passengers, and its cheap cost was inspiring narrow-gauge builders around the world. *Jennie*, a historical novel of 1958 by J. B. Snell, describes a generic Welsh slate line. It evokes very well the process of one of these railways becoming established. The rooks cease to caw in alarm when the train passes by. The villages along the route grow several times bigger, and cease to be 'farmer's villages', becoming 'quarrymen's villages'. The streets are paved and chapels are built, so the world becomes greyer. This part of north-west Wales became a percussive place, with the dynamiting of slate, and the firing of rock cannons to celebrate any new achievement of the railway.

In the 1920s, the summer service became increasingly geared to tourism as the demand for slate declined. Old photographs show trains consisting of dour quarrymen in open carriages, and happy tourists waving from the windows of more luxurious enclosed ones. Mixed traffic, you could say. In 1923, the Ffestiniog was linked to the Welsh Highland at Porthmadog. We will come to the genesis of that other railway in a minute, but let us say for now that, commercially speaking, it was always a basket case, and the Ffestiniog's involvement in it would be at great financial cost. The Welsh Highland Railway closed in 1937, and in 1939 passenger services on the Ffestiniog ceased. The railway was given a temporary boost by the war, during which the demand for Welsh slate revived. But in 1946, through services ceased. The railway was not taken into BR, just as it had escaped the Grouping of 1923. The Ministry of Transport refused to help, and the railway was abandoned, except for a short stretch that continued to be used by the quarries at Blaenau. The railway became a still life, like a photograph of a railway. Well, there was *some* stirring, in the form of grass growing on the tracks, roof timbers in the engine shed collapsing and thereafter rain falling on the locomotives.

Clare Britton puts it like this: 'The railway went to sleep but never died,' and the Ministry of Transport's indifference extended to the refusal to grant an Abandonment Order. In January of 1951, a letter written by a Leonard Heath-Humphrys suggesting re-opening the line appeared in the *Journal of the British Locomotive*. Heath-Humphrys had probably been inspired by the Talyllyn campaign, and it was this letter that deflected the abovementioned Vic Mitchell and his friend from the Talyllyn to the Ffestiniog. In *An Illustrated History of Preserved Railways* (1981), Geoffrey Body writes:

> The estimate of £2,500 for starting the service, on top of £3,500 for acquiring the line, makes interesting comparison with the sums in excess of £100,000 involved in more recent ventures. Even so, the figures were sufficiently daunting for the early 1950s and for the Ffestiniog Railway Society to need a patron in addition to enthusiast funding. Alan Pegler duly met that need and, after a series of financial and legal problems had been surmounted, acquired control of the railway on 22 June 1954.

So the dynamic Pegler goes down as the saviour of

Flying Scotsman and the Ffestiniog: the most famous engine, and the most famous preserved railway.

On 23 July 1955, a passenger service ran over the Cob – the embankment crossing the Glaslyn Peninsula at Porthmadog – to Boston Lodge, where the railway has its works. Motive power was provided by *Mary Ann*, a four-wheeled, petrol-driven First World War locomotive that had hauled men and supplies from railheads on the Western Front to the front line. In 1958, 7 miles of line were re-opened and 60,000 passengers were carried. By the time the preservationists were in a position to reopen the final stretch to Blaenau, the old line was no longer available because the Central Electricity Generating Board had built – as part of a hydro-electricity generating scheme – a reservoir engulfing that stretch. Between 1965 and 1978, therefore, volunteers built the 'Deviation': a completely new line of 2½ miles to circumvent the reservoir, creating in the process the Dduallt spiral, a formation more familiar on mountain railways, by which a track climbs by looping over itself, at which point a bridge is required.

Creating the Deviation was a matter of blasting a route as much as the laying of a line, and it involved the digging of a tunnel – the New Moelwyn Tunnel. 'Progress was measured in inches and barrels of beer emptied,' recalled one Deviationist, writing forty years later on a railway forum. If such a project had been undertaken by professional engineers, you would be impressed by the technical specifications, but since the Deviation was built largely by volunteers (with some professional supervision), you are impressed by the tightness of the psychological focus.

The 'Deviationists' of the Ffestiniog (officially the Civil Engineering Group) were the hard core, the

Praetorian Guard of the volunteers. When I was growing up in 1970s York, there was an archaeological excavation in the middle of the city. (It is now the Jorvik Centre.) Most of the diggers were volunteers – sixth-formers or university students – and they had a very distinct identity. They faced both forwards and backwards in time, being young progressives who were historically minded. Another paradox: they expressed, or offset, their intellectual interests through hard physical work, and – being essentially hippies – they had very good parties, one of which I crashed. I suspect that the Deviationists of the Ffestiniog were a similar crew. They were overseen by a paid foreman, a gruff and hirsute ex-Marine called Bunny Lewis. He had been born in Cambridge, and the Ffestiniog's online encyclopaedia, Festipedia, goes out of its way to note that he was unusual in the context of the Ffestiniog in coming from the town rather than the gown side of Cambridge.

In *The Little Wonder*, John Winton gives an example of how the middle-class Deviationists found therapy in the work. He quotes Charles Wilson, a student accountant who, one Friday in 1971, learned that he'd unjustly failed his exams 'for not sitting a paper from which I was exempted!' On the Saturday, he did a stint on the Deviation.

> 'The job in front of us was removing a tree-stump that was overhanging the shelf at Coed Dduallt – and a most intractable object that tree stump turned out to be ... We attacked that stump with mattocks, shovels and an axe, and used a winch to try and pull it over ... Well, I think as I attacked that stump I worked

off most of my frustrations and anger that I had felt on Friday – and put it to good purpose – even if we were no nearer to Blaenau, and had only a fallen tree stump to show for our pains. I really felt I had done something worthwhile.'

An example of a civil-engineering feat performed by mainly volunteers, the Deviation is on a par with the Bluebell's extension to East Grinstead. The magazine of the Ffestiniog is called *The Deviationist*. Even now that the work is complete, the men and women involved are 'ex-Deviationists', wearing the badge with pride.

Ffestiniog trains returned to Blaenau in 1982, twenty-eight years after the volunteers' takeover.

Observations

I travelled along the Ffestiniog in its Luxury Pullman Observation Carriage, which the railway built at its Boston Lodge works in 2016. It is in the Pullman livery of umber and cream and carries the Pullman crest. The use of the Pullman brand is completely legitimate. The copyright came to reside with BR's Intercity division, of which Dr John Prideaux was head. Early this century, Prideaux was chair of the Ffestiniog and Welsh Highland Railways Trust, and of the Ffestiniog Railway Company. He had become interested in the Ffestiniog when holidaying in Wales as a boy. At the time, the railway was in its still-life phase, and he had tried to move (by pushing) abandoned slate wagons at Porthmadog harbour. When the railway revived, he became a teenage volunteer, travelling from London to Wales at weekends in the back of a friend's van.

The interior of the observation car was blue plush with polished hardwoods. Each seat was a proper armchair, whereas the seats in normal Ffestiniog or Welsh Highland carriages are on a rather eighteenth-century scale. We ran over the mile-long Cob, with sandflats and seawater lagoons to the right; to the left, more sandflats and lagoons, then the mountains of Snowdonia. Coming off the Cob, we passed the low sheds of the railway's Boston Lodge works. (I visited Boston Lodge on my way back but will mention it here. As I walked with Phil Brown – 'Locomotive Manager' – amid what looked like an extended blacksmith's smithy, he said, 'We are the third biggest carriage builders in the UK, but that says more about the state of British engineering than it does about scale of our operation.' They also build and restore for other preserved railways, hence a paradox: 'You can't say this is a preserved workshop. It's a living thing.')

Beyond Boston Lodge we began our climb, and the locomotive began to announce its presence. It was a tank engine called *Lyd* and it looked about a hundred years old but was built at Boston Lodge in 2010. It bore the word 'Southern' (which must have perplexed thousands of tourists since 2010) and was liveried in Southern Railway Maunsell Green. In fact, *Lyd* is a sort of preserved-railway in-joke, in that it is a replica of a Lynton & Barnstaple Railway locomotive called *Lew*. As we climbed, *Lyd*'s cab rocked about just a couple of feet ahead of the front window of our observation car, and there was a sense of the carriage being about to crash into the engine or vice versa. It was a great piece of railway theatre, anyhow. We passed Minffordd Cemetery. There used to be a hole in the perimeter wall. The trains would stop adjacent to the hole; a coffin would be unloaded from a carriage (I hope not a wagon) and carried through. For the living, there is a connection at Minffordd to the Cambrian Coast Line.

We were now into open country, flowing along by dry stone walls. We came to Penrhyndeudraeth, built for quarrymen. The explosives factory – which supplied the military as well as quarry and mine owners – closed in 1997, and not all of the five chapels are still in use. The Ffestiniog's station in the town is called Penrhyn. Next was Rhiw Goch, the site of a farm, a manned signal box and a passing loop. Trains often wait here, but passengers can't alight. It is one of those places accessible only by rail.

We entered the National Park, and all the colours intensified: heather tones, light green grass, bright white sheep, pale blue sky, slate, flowers; then we burrowed deep into some woods. We were in the very heart of Snowdonia. The approach to Tan-y-Bwlch Station is called 'Whistling Curve'. Engine drivers whistled to alert the station manager, Bessie Jones (who moved into the station house in 1929). She would meet the trains in Welsh national costume, i.e. in a hat for which there seems to be no name, but it resembles those worn by Windy Miller in *Camberwick Green*,* or a small traffic cone. Bessie Jones retired in 1965 ... and let us return for a moment here to Vic Mitchell, that young refugee from the Talyllyn.

*I should explain, for the benefit of the youthful, that Windy Miller was, as his name implies, a miller of corn in the stop-frame puppet-animation show, *Camberwick Green*, which was broadcast on the BBC from 1966, and merits that old-fashioned word, 'charming'. Windy – whose party piece was walking nervelessly in and out of his mill while the sails were turning – dressed in medieval style, wearing a smock, neckerchief and a tall triangular hat.

He would stick with the Ffestiniog, becoming a founder director of the line. During a publicity stunt, Vic's wife greeted Ffestiniog trains in Welsh national costume. 'All went well,' Vic recalled, 'until a local newspaper reporter started interviewing her in Welsh.'

The Vale of Ffestiniog was now arrayed to our right, and everyone in the observation car was observing it. We came to a low stone house with russet hills rising behind and a wood falling away in front. This is Coed y Bleiddiau, meaning 'wood of the wolves'. Between 1925 and 1937 it was leased for ten pounds a year by Sir Granville Bantock, conductor of the Birmingham Symphony Orchestra. It is another place accessible only by train, but the trains that serve it don't necessarily stop. The drivers would drop off food or collect mail without stopping. The next tenant after Bantock was St John Philby, intelligence officer, Muslim convert and father of the spy Kim Philby. It is now lettable from the Landmark Trust, having been restored after many years' dereliction. A journalist called Tom Robbins stayed there and wrote up the experience for the *Financial Times* on 15 April 2018. He quoted Bantock's daughter: 'Our bath was the mountain pool ... no one who stayed at Coed y Bleiddiau was anything but happy there. The lovely mountains all round, the feeling of peace and being completely cut off from the civilised world, was restful and healing to the spirit.' Tom Robbins found himself fascinated by the articulated bogies of the passing Double Fairlies. 'The paradox is that at a cottage that offers silence and utter seclusion ... I've never done so much waving.'

We came to the stone halt called Dduallt (Black Hill), which marks the start of the Deviation, which involves a 300-yard-long tunnel, and the above-mentioned spiral formation by which the railway elevates itself rapidly from the reservoir – and there is a sense of flying. You're noting features of the landscape; a minute later, you're high above them.

We approached Blaenau Station, which the Ffestiniog shares with the national network in the form of the Conwy Valley Line from Llandudno. Blaenau at the height of its prosperity was a railway magnet. The LNWR and GWR were both attracted there by the slate, and competition from them in the 1880s undermined the Ffestiniog. There were once five stations in Blaenau, hideously tangled with one another, and with exasperatingly mutable names. The present station is on the site of one formerly shared by the Ffestiniog and the GWR. On arrival, about one-third of the train passengers went up to the engine, and most of them photographed the driver: a handsome middle-aged chap, with a rather theatrical amount of coal dust on his face. He beamed at everyone as he was photographed, but the town itself is a salutary place. It used to be 'the town that roofed the world'. The setting is beautiful, but there is an air of the roof having fallen in. The perimeter of the town is draped with scree of shattered slate, as if someone has started demolishing the town and is working their way towards the centre. I bought a cup of tea and a home-made jam tart from a baker's shop. The jam tart cost 40p. It would be interesting to be here on a rainy winter's day. Even the most positive accounts of Blaenau admit its tendency to dampness, the town facing west and being enclosed by mountains. Some of the surrounding quarries still do produce slate; others have become 'a mecca for extreme sport', facilitating downhill mountain biking, and Spiderman-like pursuits involving zip wires.

In the evening after my trip along the Ffestiniog, I was cycling through Porthmadog. It was a beautiful evening, the sky milky blue with a hint of pink. I heard a whistle from the direction of the station. Since no evening trains were timetabled, I rode that way to investigate. From the opposite side of the water, I watched *Lyd* going bunker-first towards a water crane. Three young men were in charge of her. *Lyd* took on water for ten minutes, with one of the men standing on the boiler to direct the flow. One of the other two rolled a cigarette while leaning out of the cab and talking to his mate on the platform.

I was staying in Tremadog, which is either a suburb of Porthmadog or a village outside it, depending on how you look at it. Tremadog is handsomer than Porthmadog, consisting of a very dignified square of grey stone houses, shops and inns, with sheer stone cliffs rising protectively behind, signifying where the coast used to be. There is also a church, which has been deconsecrated and is now the headquarters of Ffestiniog Travel, which is the best railway travel agency, and supports the Ffestiniog Railway. When I was writing a book about European sleeper trains I bought all my tickets from them. I once asked one of their clerks for tickets that would enable me to duplicate as far as possible the old Orient Express journey, and such was his historical knowledge that he asked, perfectly reasonably, 'Which Orient Express did you want to duplicate?' I was reminded of the probably apocryphal story of a homesick British serviceman who, some time in the early 1950s, asked an Anglo-Indian ticket clerk at Victoria Station, Bombay, for 'a third-class single to Leicester.'

The clerk replied, 'Midland or London Road Station, sir?'

The Welsh Highland Railway

The Beyer-Garratt

That night it rained torrentially, and I was kept awake for hours by a small waterfall rushing down the cliff just beyond my bedroom window. I didn't mind – you ought to engage with the rain in Wales – but I was tired when I returned to Porthmadog's Harbour Station for my ride on the Welsh Highland Railway.

The train would be hauled by the WHR's equivalent of a Double Fairlie: a barbarous-looking engine designed to combine power and flexibility. It was a locomotive of the Garratt type – this one built in 1909 by Beyer Peacock for the 2-foot-gauge Tasmanian Railways. Whereas the Fairlies are 'double', the Garratts look like a mash-up of three locomotives: in the centre is a long boiler and firebox, which pivots between two engine units riding on powered bogies. They seem enormous, and not just by the standards of a narrow-gauge line. The engine was 'blowing off': steam pressure had mounted to the extent that a great geyser of steam was screaming vertically out of a safety valve. Blowing off is normally a faux pas on the fireman's part, but this time it was deliberate, meant to chivvy along the platform guard to give the 'Right away': 'It's a non-too subtle reminder that we're five minutes late,' the driver shouted to me over the din.

I boarded not an observation car, but an ordinary Welsh Highland carriage. These are bigger than the ordinary ones of the Ffestiniog, because the loading gauge of the Welsh Highland line is bigger. As we set off, I placed on the table before me a lot of material on the history of the line, a tangle I have decided not to try and unpick: it's too complicated. Essentially, the route was created by a jumble of obscure railways, of which the two principal ones were the North Wales Narrow Gauge Railways (which sounds generic but was a single company), and the Porthmadog, Beddgelert and South Snowdon Railway. In 1921 they merged, to create the WHR. The line served various slate quarries and connected near Caernarvon with the London & North Western Railway at Dinas Junction, but the quarries were declining even as the line opened. In 1934, the WHR was leased to the Ffestiniog Railway company, which promoted it for tourism, but in those Depression years the line withered and fell into liquidation in 1937. We will come to the line's preservation shortly.

The Welsh Highland is the longest heritage railway in the UK. The trip from Porthmadog to Caernarfon takes two-and-a-quarter hours, involving the risk of the traveller becoming inured to the novelty of this being a picturesque preserved line, and falling into the national network mode of ennui. I myself did not become bored because I fell asleep. It happened like this.

On taking my seat, I noticed that many of the people in the carriage had figured out that the train had a buffet car; that it served such decadent treats as 'Welsh beef-burgers' and 'Jumbo Hotdogs'; that these treats, once ordered, would be delivered to your seat (which, in every case, faced a table). In the early stages of the journey, the smell of this food kept me awake even though I was tired, not having slept much the night before. Other stimuli had contributed in the same direction: the sight of the Garratt locomotive; its blowing off; then our departure from Harbour Station, which involved crossing the main road of Porthmadog. We went into a tunnel, in which the blowback from the Garratt entered the open windows, sprinkling a garnish of coal dust on the Welsh beefburgers, and replacing the breakfast smell with the – to me – more relaxing one of coal smoke. I went to sleep, only to wake up at the terminus, Caernarfon, where a new station was under construction (it will be finished by the time this book is published).

So I will describe the journey *back*.

Nutwood

Caernarvon is an impressively baronial-looking town, and not just on account of the castle. But – as at Blaenau – there are too many charity shops, and you wonder what would become of the town without the Welsh Highland Railway. It is what's called a 'Preserved Railway Town'. Unlike Blaenau, Caernarvon no longer has a station on the national network, but it would still be dependent on the Welsh Highland Railway if it did. Amongst other Preserved Railway Towns are Kidderminster in Worcestershire and Swanage in Dorset.

The journey back, then ...

At first, the stone buildings and tight stone tunnels signified the work of the North Wales Narrow Gauge Railways. The trademarks of that company would persist until Rhyd Ddu – about half the way to Porthmadog. We began to climb towards Trevan Junction, which became completely overgrown during 'the abandonment'. A 'slate trail walk' starts from here. The village of Waenfawr is half a mile off – a fairly typical situation, in that the Welsh Highland was not really designed to serve settlements of *people*. Betws Garmon is a series of sprawling villages. Thereafter, the train is in the National Park, shadowing both the Afon Gwyrfai (a river) and the

A4085. Bright green grass was giving way to grey scree under a grey sky. The grey waters of Lake Cwellyn began to be on the right, the summit of Snowdon to the left.

We passed Snowdon Ranger Halt (a request stop) from where a path leads to the summit. The path is described as 'hard'. For the indolent, the Snowdon Mountain Railway goes to the summit from the other side, taking an hour for the climb, starting at Llanberis. The Snowdon Mountain Railway – gauge 2-foot-7½ – is an old railway: it uses steam engines on some services. It is not a preserved railway as such, but it is one of 'the Great Little Trains of Wales' (which we will be coming to). At some points the line is 1-in-5: absurd for a railway, even if it is rack-and-pinion (the only one following that Swiss model in Britain). The railway opened on Easter Monday 1896, and closed on the same day after an accident: the rack failed. The line re-opened a year later. Llanberis people wanted the line because they had lost holiday trade as a result of the coming of the Welsh Highland to Rhyd Ddu, which is closer to Snowdon than Llanberis, and which for a while capitalised on the fact by calling itself 'Snowdon'. The Snowdon Mountain Railway is closed in winter and opens in March, 'weather permitting'. Apparently its ascent into clouds is enjoyed almost as much as the views it affords when the weather is fine.

We began the climb to the hamlet of Rhyd Ddu. The train then entered a forest, but breaks in the trees revealed mountain views. The track – descending now – embarked on a series of reverse curves, and Snowdon apparently jumped from one side to another. A retired couple sat opposite me (almost all the passengers were retired couples): as they listened to the sound of the writhing train, the man said to the woman, 'This is what we call flange squeal.' He had been timing the train, announcing how many minutes late it was into each station. Since the figure was never higher than three, I wondered why he bothered. We came to Beddgelert Station, which lies in a green glade above the village of that name. When it opened in 1923 Beddgelert had a goods shed, a lamp hut, a water crane, a bookstall and a waiting room – all of which proved surplus to requirements. Only the concrete base of the water tower survives from the old station.

Even on this grey day there was something magical about this wooded setting. The signals looked like rhododendrons; the engine's whistle sounded like a curlew. I saw two very well-dressed children sitting on a giant green boulder, which was odd since this was a weekday in term time. The wood seemed a place for elves or fairies ... or Rupert the Bear. I recalled watching a documentary by Terry Jones about the writer of those stories, Alfred Bedsall. He lived in a cottage at Beddgelert, and the landscape seeped into his books. Rupert often went by train. I once visited an antiquarian book shop in Stamford, where I saw for sale a bundle of Rupert monthlies from the early 1950s. The top one was called *Rupert and the Lost Railway.* I asked the shop assistant how much for the bundle. She replied, 'Oh, about eight,' and, thinking she meant eight pounds, I took a tenner from my wallet, but it turned out she meant eight hundred.

After the long, and somehow pretty, Goat Tunnel (grey stones decorated with incredibly bright green lichen), we entered the Aberglasylyn Pass, an Arthurian landscape voted in 2008 the most scenic place in the UK by National Trust members. Gradually, the landscape flattened out and became merely pleasant.

The Other Welsh Highland Railway

Coming up on our right, as we returned to Porthmadog, was a preserved station – at mention of which, a deep breath is required. The station belongs to the Welsh Highland *Heritage* Railway. Note that word 'heritage'. The Welsh Highland Railway, whose line I was riding upon, *is* a heritage railway, but the word 'heritage' is not in its title. We will call the two the WHHR and the WHR. I had first seen the WHHR station the night before while cycling along the Porthmadog Road. On one side of this lies the Porthmadog national network station; on the other is the WHHR station. (Yes, there are *three* stations in Porthmadog, Harbour Station of the Ffestiniog being the third.) The WHHR station had been closed for the evening and was fenced off. Peering towards it in the gloom, I failed to notice the word 'Heritage' on a noticeboard and took it for part of the WHR set-up. But the WHHR is a separate outfit. It began in the early 1960s as a splinter group from the Ffestiniog, committed to doing for the Welsh Highland what the Ffestiniog preservationists had done for the Ffestiniog. Legal problems prevented the group from acquiring the Welsh Highland line, so the group bought from BR the siding by which the old Welsh Highland had exchanged goods with the national network. From this siding, the group planned its next move. Meanwhile, some of the Ffestiniog people were worried about competition from the revived Welsh Highland Railway should the breakaway group be *able* to revive it. In 1987, the Ffestiniog Railway Company made a confidential offer to buy

1. This is an E2 class locomotive, as built for the London, Brighton & South Coast Railway between 1910 and 1913. The last one was scrapped in 1963, but the class lives on posthumously, in that Thomas the Tank Engine is an E2.

2. Dr Beeching re-opening the South Devon Railway in 1969. He had been invited to perform the role in a spirt of 'unconcealed irony'.

3. Tom Rolt, founder (in effect) of the preserved railway movement, a title he might not have relished.

4. Engine Number 1 on the Talyllyn Railway. The author was one of about twenty people trying to photograph it.

5. A mongrel train in the Wensleydale Railway.
(But it's a beautiful mongrel.)

6. *Typhoon* on the Romney, Hythe and Dymchurch Railway. The engines are polished with Pledge.

7. A fireman (seemingly straight from Central Casting) at Blaenau Ffestiniog station.

8. A Garratt locomotive on the Welsh Highland Railway. These engines make up in power what they lack in elegance.

9. The workshop at Aylsham on the Bure Valley Railway, Norfolk – a very un-whimsical narrow-gauge line.

10. Set dressing on the Isle of Wight Steam Railway.

11. A waxwork of Colonel Stephens, the builder (in spite of his bristling appearance) of some of Britain's prettiest and sleepiest lines.

12. A stuffed bull's head in a mocked-up freight train on the Mid-Hants Railway. Note amusing chalk inscription.

13. A locomotive called *Doll* in Christmassy mood on the Leighton Buzzard Narrow-Gauge Railway.

14. A mellow lounge car on the Strathspey Railway.

15. The author's view (one late-March day) from a carriage of the Wells & Walsingham Light Railway.

16. Coal wagons on the Great Central Railway. The Morris Minor is probably of about the same vintage, as well as the same colour.

17. A hot afternoon on the Kent & East Sussex Railway.

18. Blood and custard carriages harmonising with the landscape of the North Yorkshire Moors Railway.

19. A BR Standard Class 4 on the gritty East Lancashire Railway.

20. The station shop at Embsay on the Embsay & Bolton Abbey Railway. The shopkeeper was doing her knitting at the time.

21. A scene of dereliction (or potential) at Bolton Abbey station.

22. ... And what more could one require?

23. Seat moquettes are now designed according to principles of geometry and colour theory. In the past, they were just meant to look pretty.

24. A Warship-class diesel on the Severn Valley Railway. The Warships are definitely among the interesting diesels.

25. A Class 26 at Dereham on the North Norfolk Railway. 'It's as much part of railway history as steam,' the driver had just informed the author.

26. A Derby Lightweight (with 'speed whiskers') on the Ecclesbourne Valley Railway.

the entire Welsh Highland trackbed from the Receiver* in order to prevent the reopening of the line. When news of this move leaked out, the Ffestiniog company was cast in a bad light.

Personnel changes within the company led to a new, more noble strategy. The Ffestiniog Company would aim to buy and re-open the Welsh Highland line, which is what the splinter group also wanted to do. Battle commenced and, in 1995, the Ffestiniog people won. With funding from the Millennium Commission, the Welsh Government and generous private supporters, they reopened the WHR over its full length in 2012.

So the exchange siding became the permanent home of the WHHR. The WHHR website advertises a 'small, friendly railway', and I am told it *is* friendly to its visitors, but I was also told that some of the bitterness towards the WHR lingers, and there can be 'acrimony at board level'.

The website of the WHHR urges visitors to 'look out of the window and see the mountains and towering cliffs', but the beautiful scenery is some distance away, and I was put in mind of a prisoner wistfully describing a view from his cell.

The Great Little Trains of Wales

The Great Little Trains Differentiated

Despite the tension between them, the WHHR and the WHR are both members of the marketing scheme created in 1970, the Great Little Trains of Wales. Eleven

*It was remaining in receivership that protected the line from development.

railways are members, and we have encompassed seven so far in this book: the Talyllyn, the Ffestiniog, the Welsh Highland, the Welsh Highland Heritage Railway, the Snowdon Mountain Railway, the Llanberis Lake Railway and the Fairbourne Railway. The others are the Bala Lake Railway, the Vale of Rheidol Railway, the Brecon Mountain Railway and the Welshpool & Llanfair Light Railway. Not all are as deeply rooted in Welsh history as the Talyllyn or Ffestiniog.

The Bala Lake Railway was set up in 1972 as an avowedly tourist railway along a stretch of what had been the standard-gauge Ruabon to Barmouth line, of which more in a moment. The Brecon Mountain Railway also runs along a stretch of what had been a standard-gauge line (the Brecon & Merthyr Railway). It was opened in 1980 by a businessman called Tony Hills, who had accumulated a number of locomotives at the Llanberis Lake Railway. In *British Railway Enthusiasm*, Ian Carter classifies the Brecon Mountain Railway, along with the Snowdon Mountain Railway, as examples of 'preserved' railways run as 'more or less straightforward commercial investments'. Few standard-gauge preserved lines are run on that basis: one is the Dartmouth Steam Railway (Paignton to Kingswear in Devon). The majority are run 'not for profit'.

(Another part of the Ruabon–Barmouth line, which closed to passengers in 1965, is occupied by the Llangollen Railway, which is a preserved Welsh *standard*-gauge line, therefore not one of the Great Little Trains of Wales at all. The principal station is advertised by a sign reading, 'Llangollen, GWR', not 'Llangollen Preserved Railway' and, since the station is in old GWR colours, the clear message is that the last seventy years haven't happened. At its workshops the railway is rebuilding – *Jurassic Park*-like – some extinct classes of locomotive, including an ex-GWR Grange class. The Llangollen is the longest, at 10 miles, of the five standard-gauge lines preserved in Wales. The others are the Barry Tourist Railway, the Gwili Railway, the Llanelli and Mynydd Mawr Railway, the Pontypool and Blaenavon.)

The Brecon and Snowdon Mountain Railways are in tourist hotspots, which is why they can be run as commercial concerns. The Vale of Rheidol Railway also benefits from a spectacular location, and it is entangled with the Brecon Mountain Railway, as will be apparent after a little history. This 1-foot-11½-inch line was opened in 1902, running 12 miles from Devil's Bridge, 600 feet up in the Cambrian Mountains, to Aberystwyth. The original aim was to carry lead ore and timber to the sea, but the emphasis soon switched to tourism on what was a very scenic route. In 1923, the Vale of Rheidol line was taken over by the GWR, who built a new station for it beside the standard-gauge one in Aberystwyth and promoted the tourist traffic. BR would do the same, allowing steam to continue on the line after it had disappeared everywhere else in 1968. The boxy tank engines of the line were the only steam engines painted in 'rail blue' and with the double arrow logo on the side: a case of mutton dressed as lamb. BR borrowed some gimmicks from the preserved railways, laying on Santa Specials. There were also (and these were possibly unique to the Vale of Rheidol) outdoor theatrical entertainments called 'Wild West Attacks'. Edwardian photographs of the line do show a Wild West aspect: thin men lounging about on the platform at Devil's Bridge (which would be a good name for a Western) next to the tin-shack station building.

In 1989 the line was sold to Tony Hills and his business partner, owners of the Brecon Mountain Railway, so becoming the first part of BR to be privatised – or so I thought, but then I read in a book called *The Vale of Rheidol Railway* by Hugh Ballantyne that the line was actually 'the second part of the nationalised system to be sold to a private buyer'. He does not say what the first was. In 1996, Tony Hills departed, and the line was then run by a charitable trust. For the next twenty years, the line was run by this trust and without volunteers, which is unusual for a preserved railway, but then again, is it preserved? After all, it never closed.

The Welshpool & Llanfair is not so auspiciously sited. On my way by car to Porthmadog, I had just crossed into Wales late on a Sunday afternoon, and I was aiming towards the sun setting over west Wales, when it seemed I was being tracked by another, smaller sun on the other side of a hedgerow. It was a couple of seconds before the image resolved itself into sunlight glinting off the burnished copper cap of a tank engine chimney. The hedge dipped, revealing the driver – a young, Edwardian-looking man – leaning out of the cab. He just had time to wave at me before the hedge intervened again, and it seemed as if he was just driving through farmers' fields.

I beat the train to its station, which was Llanfair Caereinion, westerly terminus of the Welshpool & Llanfair Light Railway. This railway was opened in 1901 under the provisions of the 1896 Act and was operated by a company called Cambrian Railways. The aim was to link farms to Welshpool, and the line used to run through the town towards exchange sidings a little way north of the old Welshpool Station, which resembles a French chateau, and which now, according to a mid-Wales tourism website, 'offers a unique opportunity for both the holidaymaker and the local visitor to shop in one of the most elegant listed buildings in the area'. The station was killed by development of the A483 (the Welshpool bypass) in the early 1990s, which required the building of a new station to the south. This new station has a single island platform, and a passenger 'help point'.

There's a photograph online of one of the engines going through the town in the 1940s, as Welsh Highlands do once more in Porthmadog. Passers-by are looking on with surprise, as if they'd never got used to the idea of a locomotive in the streets, and the engine looks big and dangerous, even though this was a 2-foot-6-inch-gauge line. (The W&L engines clanged a bell continuously as they rolled through the town.)

The line was acquired by the Great Western in the 1923 Grouping, and the loco I'd overtaken was in Great Western green, and its coaches were GW chocolate and cream. The old W&L was not a success, having too few visiting tourists to counteract road competition. It lost its passenger services in 1931, closed in 1956 and was reopened by enthusiasts in 1963, although it no longer runs into Welshpool: it stops at Raven Square on the edge of the town, but there are tentative plans for it to connect to the new Welshpool Station by a route that would avoid the streets of the town.

At Llanfair, I stepped onto the platform and greeted the driver. There hadn't been many people on the train, and soon we had the station to ourselves; our shadows were long on the platform. I asked the driver which other lines had a 2-foot-6-inch gauge.

'Whipsnade Zoo?' he said, and I was reminded that I used to take my children to Whipsnade precisely to ride on that line, formally the Great Whipsnade Railway.

'Also ... the Zillertalbahn in Austria,' said the driver. 'We have four coaches from there.'

Rolt's Ghost and Jennie

The marketing of those railways as The Great Little Trains of Wales has been a great success. Even so – or perhaps for that very reason – I don't think Tom Rolt would have approved. We have already noted his prejudice against tourism in Wales. It was perhaps first expressed in one of his earliest books, a collection of ghost stories called *Sleep No More*, published in 1947. One of the stories, 'Cwm Garon', is set on the Welsh borders. *Cwm* means a mountain basin. A man called John Carfax travels there by changing from the express from London to a branch line. The train gradually empties out as Carfax approaches his destination: all the 'black-gaitered farmers and plump, basket-laden wives' had gone, 'but still he seemed to smell sheep-dip and carbolic, to hear the lilt of their border speech, and to see the lithe Welsh sheep-dog which had sat between his master's legs, regarding him with wall-eyed suspicion.'

Things get progressively more ominous. Carfax meets a gloomy folklorist at an inn: 'For centuries this dark power has been, as it were, dammed up in this valley until it has soaked into the very stones of the place ... Outside interference has an effect upon it like that of a stone flung into a still pool. That's why Cwm Garon and its people have always implacably resisted intrusion.' I will not give away the ending. It seems possible that Rolt's position is explained by a familiar psychological inversion. He disliked offcomers because he feared that he himself was one.

The other book I think of when I read those perky words, 'Great Little Trains of Wales', is J. B. Snell's novel, *Jennie*, which describes the rise and fall of a narrow-gauge slate line, from the mid-nineteenth to the mid-twentieth century. Snell's tone is unsentimental. There is nothing cute about his railway, and he sees the social divisions it promotes. The villagers look down on the navvies who come to build the railway, and they're glad when they leave. The railwaymen are a cut above the quarrymen. The carriages that take the quarrymen to and from work are more primitive than the carriages used for the public passenger services. But sometimes the quarrymen travel on the public passenger services, for instance on Saturday night when they return to the hill villages after a night's drinking in the port from where the slate is exported. The quarrymen board the train in leisurely manner, ordering the crew to wait beyond the departure time, because some of their mates are still on their way to the station. Here is the eponymous three-cylinder tank engine, *Jennie*, as she turns her back on the sea and heads into the hills:

> The long train wound slowly up the hill, the yellow lamplight from the coaches sending pale shadows dancing through the bare trees ... Everyone in every coach was singing, and in each coach a different song was being sung. It made the night hideous, and by the

light of the headlamp Owen [the driver] saw a large tawny owl give him a long, jaundiced look before spreading his wings and flying off clumsily into the darkness.

In the early 1920s the managers of the railway decide to 'go after the holiday traffic', and a pattern is set whereby the railway has a boom time every summer but business in the rest of the year gets gradually worse. At first, the locals can't believe that the tourists would want to go to the hillside locations they themselves have taken for granted all their lives. It seems they even want to see the quarries. To the locals this is like people wanting to go 'nowhere'. When the railway advertises the places along the line with brightly coloured posters, the locals hardly recognise their own homes. (But here is an important fact about slate mining: it scarred but did not destroy the landscape.) In summer, the slate traffic runs at night, the days being taken up with tourist trains. The managers of the line sometimes have to press the workmen's carriages into service for the tourists but, stricken by conscience, they characterise them honestly as 'Fourth Class' and charge a very cheap fare.

In the mid-1930s, the quarry gradually goes bankrupt: 'it was a scandal and an outrage, and all the fault of the quarry company, the county council, the government, the banks, or even sometimes the Americans. Nobody thought of blaming the people who decided they preferred tile roofs, although of course they were the real villains.' Tourism declines in the years of the Depression, and a bus company competes for the local passenger trade. The bus men are chancers, as in *The Titfield Thunderbolt*, but *Jennie* is no Ealing Comedy, and no preservationists come along to save the line. My edition of the book is illustrated by G. K. Sewell, and the last image is a scene of railway dereliction, skilfully drawn.

Why did Snell allow his fictional line to die? He published just as the Ffestiniog was beginning to revive, and Snell himself was a distinguished preservationist who, as a young volunteer, had fired the first train to be run by the Talyllyn Railway Preservation Society before going up to Oxford. Later, he was chairman of the North Norfolk Railway and General Manager of the Romney, Hythe & Dymchurch. But there is an aesthetic rightness about the dwindling of the line in his novel; a happy ending would have seemed crass. Perhaps Snell was simply experimenting with fatalistic thoughts similar to those expressed by Rolt who, in *Railway Adventure*, conceded that there was a case for letting any apparently moribund institution die because preservation would merely succeed in 'embalming a corpse from which the spirit has flown'. But Rolt felt himself exempt from that consideration by the age in which he lived: 'a Dark Age, where evils of unprecedented power threaten the whole world. And when the house is afire the instinct to save something of the blaze is too strong to be denied.'

But I think Rolt would have been a preservationist whatever era he had lived in.

5

Railways in the Landscape

Moss Upon the Sleepers

The relationship between railways and the countryside in Britain has been harmonious since about the 1870s. Before then, railways were seen as agents of urbanisation, extending towns, intensifying inner-city slums by depositing there the poor (and disenfranchised) rent-payers who had been evicted to make way for the new lines, and generally projecting the noise and chaos of industry into the countryside. The squirearchy protested that sheep and cattle would be made barren, crops would fail. Landowners would agree – at a price – to a railway incursion, but they insisted that the trains be hidden in cuttings, which is why, today, you can usually see a road from a train but not the other way around.

We have noted William Wordsworth's objection to the Kendal & Windermere Railway. But let's do a jump-cut to 2005 and the *Guardian's* obituary of Dr Peter Beet, rail enthusiast and founder of Steamtown at Carnforth. 'As Peter's remains were lowered into the grave in Kendal, perhaps by chance, yet right on cue, a train whistle sounded in the distance.' What accounts for the transition? People became accustomed to trains, and railways became an honorary part of the landscape, welcomed in a way that cars never would be, except perhaps during the 1920s when inn-keepers in villages by-passed by the local railways lured motorists with a carriage lamp emblazoned with the AA logo, and it was possible to speak of 'the romance of the road' – which was really the romance of the *empty* road, therefore short-lived.

According to Ian Carter in *Railways and Culture in Britain: the Epitome of Modernity*, our romanticisation of the country branch line is traceable to a volume published in 1893: *The Delectable Duchy*, by the Cornish folklorist, Arthur Quiller-Couch. It contains a short story called 'Cuckoo Valley Railway'.

This fictional railway line was opened in 1834, serving Tregarrick (a stand-in for Dartmoor) and going from there to 'nowhere in particular', carrying ore, clay, coal and people. Some decades later, it has one engine, satirically called *The Wonder of the Age*, and nature is covering 'the scar of 1834'.

> We climbed on board, gave a loud whistle, and jolted off. Far down on our right, the river shone between the trees, and those trees, encroaching on the track, almost joined their branches above us. Ahead, the moss that grew upon the sleepers gave the line the appearance of a green glade, and the grasses, starred with golden-rod and mallow, grew tall to the very edge of the rails.

Carter contends that 'Moss upon the sleepers' is the first depiction of a railway as endearing underdog. So it's at the base of the romantic Betjemanesque edifice. *The Railway Children* (1906) and Edward Thomas's poem 'Adlestrop' (written in 1914, published in 1917) also placed the railway at the centre of a country idyll. 'Many of the steam trains run privately will survive not through people playing trains,' wrote David St John Thomas in *The Country Railway* (1976), 'but because young and old come to enjoy the scenery they unlock, reaching places inaccessible by road or with inadequate parking facilities, winding along shelves cut into mountain sides, jumping over ravines, intimately immersed in their landscape.'

The most unprofitable lines were the country branches, or the lines to the seaside with their seasonal traffic, so these were the ones Beeching tended to close; they are in turn usually the ones that have been reopened, but not necessarily as preserved railways. Local authorities and Sustrans, the sustainable transport charity, have brought back 4,500 miles of closed-down trackbed as public walks or cycle trails.

The country branch lines are well evoked by St John Thomas. He describes stations with cattle pens alongside, so that they looked more like farms than stations; nesting boxes hanging off the signal box, and a horseshoe for luck. Until the early 1950s, a passenger train on these lines might also include a couple of freight vans, and the guard might share his van with seed potatoes, flowers, pigeons, poultry, milk (but not cream; that would not keep).

The railway writing I most enjoy describes the country lines. There is this, from *The Thirty-Nine Steps* by John Buchan:

> About five o'clock the carriage emptied, and I was left alone as I had hoped. I got out at the next station, a little place whose name I scarcely noted, set right in the heart of a bog ... An old station-master was

digging in his garden, and with his spade over his shoulder sauntered to the train, took charge of a parcel and went back to his potatoes. A child of ten received my ticket, and I emerged on a white road that straggled over the brown moor.

One of the most intensely bucolic railways in fiction is *The Forest of Boland Light Railway* (1955), by the naturalist, illustrator and author Denys Watkins-Pitchford, whose pen name was BB. The story is a bit weedy, although the illustrations are charming. This and all BB's books begin with the same moving inscription, which he had once seen written on a gravestone:

The wonder of the world, the beauty and the
Power, the shapes of things, their colours, lights,
and shades; these I saw.
Look ye also while life lasts.

The Boland is a very light railway indeed. The gnomes who built it 'did not lay more than six inches of track a day.' John Hadfield's novel of 1959, *Love on a Branch Line*, is a lament for a country branch while also being an erotic extravaganza, concerning as it does an eccentric, elderly rail enthusiast* who presides over a crumbling branch line with his three nubile daughters.

What appealed about the country branch was the sense of decay. There is this remembrance of country stations from *London General*, a series of Edwardian recollections published by Frederick Willis in 1954:

*Despite having lost his legs in a railway accident.

> We liked to have a drink in the buffet when one was available, and become heavily facetious with the waitress. Those country railway buffets were like pretentious Victorian refreshment rooms that had died fifty years earlier and were then embalmed; there were large, smeary mirrors with startling stains of damp which looked like maps of unknown countries, cold, marble tables stained with generations of tea-cups, steaming urns round which grubby tea-cloths were eternally drying, and round glass cases on the counter, one containing Banbury cakes, relics of former civilisation, and the other ham sandwiches, coldly lying in state and embellished with a sprig of parsley.

Willis remarks the curious smell of booking hall: 'steam, eau-de-cologne and dust'.

A really sophisticated preserved railway might attempt to create or 'curate' this haunted-house mood. Most preserved railway premises look recently tidied, and they might smell of cleaning fluids. I have not yet seen a preserved station that is deliberately 'distressed', like Dennis Severs' house in Spitalfields, London, an eighteenth-century house presented as mouldering, with dripping candles, half-eaten bread and orange peel on the tables. I would like to see a preserved country station waiting room with horsehair bursting through the leather of the benches, soot stains around the fire, inadequate, seething gaslight, and an unemployed actor playing the part of the waiting-room bore – some authentically smelly yokel, banging on about potato blight while emitting acrid pipe smoke. But perhaps that would be overkill. The fact is that most preserved lines are about the countryside as well as the railway. Indeed, as Patrick Wright observes in *Living in an Old Country*, 'Isn't the very look of rural England now historically identified with the "historical" itself?'

Great Western Scenes

Didcot

It seems to me that the railway most associated with the countryside was the old Great Western.

The main London suburban services ran to the north, east or south, so the GWR did little of that drab work. On the other hand, the GWR has more index entries in *The Country Railway* than any other company. The action of *The Titfield Thunderbolt* takes place on GWR metals, and most of the doomed stations mentioned in Flanders and Swann's song of 1963, 'The Slow Train', are GWR ones. The GWR had the best weather, as its publicity material constantly asserted, to the point of implying that, meteorologically speaking, it was a toss-up between 'the Cornish Riviera' (a place invented by the Edwardian GWR) and the actual Riviera. The GWR was known for its Ramblers' Specials and Mystery Trains into the countryside. The definitive country branch train was perhaps a Great Western Prairie-class tank with a short rake of chocolate and cream coaches.

'The popularity of the GWR is one of the peculiarities of railway preservation,' writes Denis Dunstone in *For the Love of Trains*:

> Perhaps it was due to the association with holidays, or the charming branchlines in picturesque countryside

> ... Of preserved railways there are more seriously authentic GW lines than any other, and there are more GWR-built steam engines preserved (some 130) than any other (the LMS is runner up with about 70). The Didcot Railway Centre has come to reflect this unique status.

The Didcot Railway Centre is accessed via Didcot Parkway station – an important junction on the Great Western main line, and well named, since it is surrounded by car parks, all of which were full when I visited.* It occurred to me that Beeching's idea of people driving to a 'railhead' had belatedly come true.

The Railway Centre was established in 1967 by the Great Western Society, which had been founded in 1961 by a group of schoolboys. It is an enthusiasts' venture, run by volunteers, whereas Steam, the museum of the Great Western at Swindon, is run by paid staff. Steam shares the site of the old Great Western works ('the factory', where the GWR built its engines) with the Great Western Designer Outlet. The Didcot Railway Centre is sited on the ashy hinterland of an old GWR engine shed. The trains of the modern-day operator – also called GWR – thunder past, but they are in a parallel universe. To access the Centre you go through the barriers of the modern station (staff will open them for you) and walk through an underpass, in which you are acclimatised for your trip back in time by a series of old GWR posters: 'Devonport: Delightful Excursions; Frequent Naval and

*The multi-storey one alongside the station is a conversion of the old GWR provender store, where hay was kept for the railway horses.

Military Manoeuvres'. 'Criccieth: Ideal Summer and Winter Climate'. I walked past an assortment of Mr Churchward's elegant green engines, and entered the Museum, where I met a very helpful man whose name I did not catch, but whom I will call the Curator.

The Curator told me of the great loyalty the GWR inspired, which persisted long after it become the Western Region of BR. 'They were still using GWR stationery at Paddington in the 1970s.' I suggested that the appeal of the railway was partly down to its fogeyish eccentricity: sticking with the broad gauge and lower quadrant signals after other companies had decided they were illogical; kitting drivers out in white fustian, and positioning them on the right- hand side of a cab; having engine numbers in brass relief rather than merely painted; signs reading 'Tickets will be shewn' or (in Gentlemen's lavatories) 'Please adjust your dress.'

The Curator said there was something in that, 'But it was also the way they looked after their staff. They paid well, and there were lots of clubs and societies.' He showed me to a glass case that featured a poster advertising 'the GWR Social and Educational Union, Grand Floral Fete, August 22nd 1936.' It was held at Botley Recreation Ground, Oxford. 'The proceedings will be opened at 2 p.m. by Sir William James Thomas Bt, Director of the GWR.' There was a surprisingly high erotic content: 'Magnificent Display by Fifty Ladies (Paddington), arrayed under the auspices of the League of Health and Beauty ... Parades of Bathing and Sports Girls.' There was a 'Beauty Competition (For Ladies)' and 'Competition for Handsomest Gent'. The grand finale would be a 'Great display of fireworks.' All proceeds would go to 'the GWR Helping Hand Fund.'

The GWR operated its many rural branches with enthusiasm and, Flanders and Swann notwithstanding, speed. It was in the interests of speed and efficiency that, in the early twentieth century, the Railway pioneered the use of railcars, which were self-propelled carriages (they looked, accordingly, like headless chickens) incorporating a steam engine with a vertical boiler. They did away with the need for the engine to run around its train and, being quick to stop and start, allowed more stops to be made, often at halts that were little more than hutches in the middle of fields.

The Curator walked with me around to a carriage shed to see Railmotor Number 93, built in 1903, acquired by the GWR Society in 1970, and restored a few years ago with Heritage Lottery funding. Then – and this was a GWR speciality – came autocoaches: a carriage with a driving cab at one end, but no engine, since it was designed to work in partnership with a small tank engine coupled at the other end, and when that engine *was* coupled on, the double act became an auto*train*. Autotrains can be seen on the Severn Valley Railway, the Llangollen, the Bodmin & Wenford and the South Devon ... and there's one at Didcot, which was waiting at the platform on the demonstration line as we walked away from the viewing of Railmotor 93. I took a ride on Didcot's autotrain, and I wished I could have gone further, but the Didcot 'branch line' affords only a three-minute trip.

Passengers on these trains would have been impressed (if they were not anti-social) by the banishment of compartments in favour of open carriages. 'Part of the ritual of travelling by country train', writes David St John Thomas in *The Country Railway*,

> was to see and hear one's fellows airing their preoccupations of the moment, listen to the guard calling out the stations ... or shouted in ecstasy that a field was plastered with mushrooms. Autotrains were one class only; you chose the best seat and if Lord Whatsit came after you he would take second choice – which seemed unusual then.

When I disembarked from the autotrain, the Curator took me over to a small carriage shed for a look at a descendent of Number 93 – another example of a GWR railcar. This was one of the GWR's streamlined, diesel 'Flying Bananas', which operated successfully between the 1930s and 1960s, bringing a Futuristic touch – especially when skimming along at 70 mph – to West Country rural scenes. Their curvature is indeed somewhat banana-like, and they are approximately the colour of a banana sundae if you drizzled chocolate sauce over it. Aside from the one at Didcot, which occasionally runs, there's one at the National Railway Museum, and one under restoration on the Kent & East Sussex Railway.

By now, afternoon was turning to evening. The darkening of the sky was being assisted by the smoke and steam of a dozen engines fired up at the mouth of the engine shed. It seemed that everyone remaining at the Centre except me carried an expensive camera, and I got the impression that these photographers wanted me out of the way, so they could begin communing with the engines that had been laid on specially for this shoot. I felt – wrongly, of course – that something intimate and furtive was afoot, like those sessions (which I believe occur) where groups of men are invited to come along and photograph a glamour model. So I walked back along the time tunnel to the traffic roar of modern Didcot.

The Swindon & Cricklade Railway

The Swindon & Cricklade Railway is not easy to find. It doesn't go into either Swindon or the village of Cricklade, which you'd think it would, given the name, and one of its three stations, Blunsdon, has lately become enmeshed in a big new estate of executive homes. But I found Blunsdon Station eventually, and even in the scruffy station yard I knew I would like the S&C.

It was as if the housing estate had never existed. I was in a sort of ashy railway colony created in a glade of trees. Under a broken, barn-like structure was most of a Merchant Navy-class locomotive. There were antiquated cranes, collapsing brake vans, crocked wheelbarrows and lawnmowers, old phone boxes and various shacks and sheds in GWR light and dark stone. I was reminded of the dilapidated but compelling railway lands around Howrah Station in Calcutta, perhaps because I had arrived at Cricklade on a day of tremendous heat and hazy orange sunlight. I bought a cup of tea in a café incorporated into an antiquated carriage of the Norwegian railways, and studied a guide to the line, which was only a free leaflet, not the glossy booklet ('That'll be five pounds, please') of the bigger lines. The S&C runs over just 2½ miles; it has no paid staff and only 600 members, but a high percentage of these are active, and the line is proud of having a five-road engine shed – containing various more or less rusty works in progress – at its northern terminus of Hayes Knoll, from which it will soon be stretching out a further 2 miles to reach Cricklade.

The S&C seems profoundly peaceful and rural, the ideal place to recuperate after the orbital roads of Swindon. It does not preserve part of a branch line, however, but part of the old through line which ran from Cheltenham to Andover via Swindon Town Station (as opposed to just Swindon), which no longer exists. The line was the work of the Midland & South Western Junction Railway, which was incorporated into the Great Western at the 1923 Grouping.

During a short ride in a short and largely empty train to Hayes Knoll, I realised that Mark 1 seat cushions could not cope with global warming: they were almost too hot to sit on. I got talking to the young train guard, whose name was Luke. He told me of the Railway's ambitious plans, and he was proud of being part owner of an early DMU, which he pointed out to me as we rolled past it. He told me it had recently been attacked by vandals; there'd also (a few years before) been an arson attack in the engine shed. 'That sort of thing slows us down considerably,' he said.

I asked why he thought people vandalised preserved railways.

'Because they don't *understand*,' he said.

Preserved railways are often vandalised, and perhaps this is just because they are unattended at night, and a railway carriage is an easy thing to vandalise, what with all those windows. On a couple of occasions when I've been on guided tours of depots, it has been explained to me that a carriage under repair is not so much being restored to its original condition as being restored to its condition prior to a recent vandalisation. Perhaps there is a provoking whimsicality about preserved railways, or perhaps the vandals are jealous of the alternative worlds the preservationists have created.

I mentioned to Luke that I'd visited Didcot, and he said, 'It's very interesting there. If they think you like the Southern [Railway], they might not take you on. You have to prove your loyalty to the Great Western, and they'll interview you to make sure of it.'

'But surely *you're* loyal to it?' I said. 'After all, your line came under the Great Western.'

'Yes,' he said, 'but only in 1923. Before that, we were the Midland & South Western Junction, which the GWR regarded as a rival.'

So I suppose that Luke might be regarded as a quisling at Didcot.* (The GWR patronisingly referred to the M&SWJ as the 'Tiddly Dyke' railway, and the Swindon & Cricklade has modestly adopted that as the title of its magazine.)

At Hayes Knoll, the S&C has built a valanced canopy that places the platform in deep shade: the effect, together with the old luggage barrows, the sparsity of people and the loud birdsong, was very poetic: a dusty summer gloom of the kind you sometimes find in old barns and cricket pavilions. I sat there between trains, thinking of a Walter de La Mare poem, 'The Railway Junction'. Verses two and three convey the salutary loneliness of a quiet station in summer:

How still it is; the signal light
At set of sun shines palely green;
A thrush sings; other sound there's none,
Nor traveller to be seen –

*From my own experience of Didcot, though, I doubt it.

Where late there was a throng. And now,
In peace awhile, I sit alone;
Though soon, at the appointed hour,
I shall myself be gone.

The West Somerset Railway

The West Somerset Railway *ought* to begin at the important railway junction of Taunton. It runs along the bulk of what was a branch line from Taunton to the seaside town of Minehead, but the WSR trains don't usually run into Taunton. Its southerly terminus is, rather, at nearby Bishops Lydeard, where I arrived just in time for the mid-morning departure for Minehead. I actually noticed the man at the regulator of the engine – *Raveningham Hall* – stay his hand as I ran along the platform, towards a door held open for me by a guard.

'I don't have a ticket,' I gasped.

'Jump aboard,' he said. 'You can pay on the train.'

On the train, I began looking at the lightly wooded landscape to my left, trying to spot Combe Florey, the Georgian house in which Evelyn Waugh had lived. I couldn't see it amid the trees, which is probably just as he would have liked. I can find no reference to the local branch line in his letters and diaries. When he had visitors, he seems to have collected them by car from Taunton. I then took out some reading matter about the WSR.

The original West Somerset Railway was opened in 1862; in 1874, it was extended to Minehead. In 1876 it became part of the Great Western, and in 1882 the whole line was converted from broad gauge to standard in twenty-four hours. The platforms were extended in

Edwardian times, a reflection of the line's popularity with tourists. It made money in summer, lost it in winter. Business was boosted in 1962 when Billy Butlin opened a holiday camp at Minehead. Butlin was keen on railways and operated his camps in conjunction with them. Some had their own branch line. In 1964, he brought two locomotives to the Minehead camp for children to play on, thus saving them from the breakers. One was an LMS streamliner of the Princess Coronation Class, *Duchess of Hamilton*, now in the National Railway Museum. Butlin acquired another engine of the same class, *Duchess of Sutherland*, and took it to his camp at Ayr. It is now based at the Midland Railway Butterley, a line that does for the Midland what the Didcot Centre does for the GWR. Perhaps Butlin wanted to rescue engines, having endured – like so many steam locomotives – a miserable time at Barry in Wales. He was apparently inspired to found holiday camps by having had a rain-sodden childhood holiday on Barry Island, during which his family were locked out of their guest house all day.

Slowly, the line was strangled by car use; it closed in 1971. A poster announcing the closure is displayed at Bishops Lydeard Station:

LAST TRAIN
Saturday 2 January 1971
Minehead – Taunton – Minehead
Depart 7.30 p.m.
Approximate arrival 10.05 p.m.
Tickets £1, inclusive of sherry reception at 6.45 p.m.
at the Beach Hotel.
Commemorative booklet and souvenir ticket.

Sponsored by Minehead and District Round Table.
All proceeds to local charities.

But there were already plans to revive the line, and Somerset County Council had secured the track, which it leased to a newly formed West Somerset Railway Company. The line was gradually re-opened between 1976 and 1979. The original plan was to run trains through to Taunton, and by 1979 the line had reached the junction at Norton Fitzwarren, just west of Taunton. But the final push on Taunton was blocked, strangely enough, by the National Union of Railwaymen. The union contended that if the WSR ran into Taunton it would undermine the jobs of NUR members who worked as bus drivers between Taunton and Minehead. The NUR 'blacked' the Railway, refusing to run a supporting transport service or to signal the WSR trains into Taunton. It has been speculated that its opposition owed something to resentment of private railway operation per se, and the WSR *had* originally intended to be more than a tourist railway: it had sought to run a daily diesel commuter service, and freight services. In any case, WSR trains still do not reach Taunton, although occasional specials do run onto the WSR from the national network via Taunton.

Today, the line is oriented to tourism. It carries over 200,000 people every year, has over 1,000 active volunteers and fifty employees. It is one of those super-charged preserved lines. It has an engine shed, at Williton, that overhauls engines for the WSR and other lines; and there is an apprenticeship scheme. At 23 miles, it's the longest standard-gauge independent railway in the UK – and long enough to be both a country line and a line along the sea.

As you leave Bishops Lydeard, the Brendon Hills are to the left, the Quantocks to the right, and nothing disrupts the branch-line reverie. On the weakly sunny day on which I travelled hay had been gathered in old-fashioned-looking stooks in the adjacent fields. You can see why the line was the setting for the ITV children's TV series of the late 1970s, *The Flockton Flyer*, the star of which was a GWR pannier tank, number 6412, which was the first engine to run on the re-opened line, and is now on the South Devon Railway, where it continues its showbusiness career, playing the part of 'Duck' on Thomas Days. In the series, the engine enabled the Carter family to revive a closed-down railway, which was a moral improvement for them, since they had previously run a petrol station. The show was not spared the nit-picking of some enthusiasts, who pointed out that locomotive names (unlike train names) were not usually prefaced with the definite article; so the programme was called *Flockton Flyer* for the second series.

Approaching Watchet the West Somerset line runs along the cliff top, at one point 10 feet from the edge, which, as the pocket guide to the line anxiously notes, is 'closer than when the line was built!' Watchet harbour on the Severn was once surrounded by a mass of sidings for coal, timber and iron ore. The harbour closed to commercial shipping in 1993; today it's a marina. The line then goes inland to Washford, rejoining the sea at Blue Anchor, a settlement created by the railway for holiday-makers. Dunster Station, serving 'the most visited village on Exmoor', was the model for Hornby Dublo's branch-line station (a compact, brick building with a gable roof and two tall chimneys).

Minehead is a rangy station, in light and dark stone,

with a double-sided valanced canopy. It is right in the centre of Minehead, being at the end of the main street which, tellingly, is called The Avenue. Minehead was always genteel. As at Porthmadog, the station seems so integral to the town – everybody seems to be either going to or coming from it – that you can't believe it is merely a preserved station. It is not so much timeslip as a kind of landslide.

The Somerset & Dorset Joint Railway: a Diversion

On the way back to Bishops Lydeard, I alighted at Washford, which was quiet, the long, slightly grass-grown platform haunted by the ghosts of Edwardian day trippers. A congenial volunteer was listening to *Test Match Special* in the station shop. I asked him why the station was painted green, the colour of the Southern Railway, whereas all the others were in GWR colours.

He gave the slightly guilty smile of someone who sees the chance to tell a story they like telling and have often told before. He indicated the building on the other side of the line, which had once been the principal station building, and was now a Museum belonging to an organisation dedicated to an entirely different railway: the Somerset & Dorset Railway Trust. So here was a memory within a memory: an enterprise celebrating an entirely different line to the West Somerset, and one frequently commemorated and lamented. In his book, *Somerset Steam,* Michael Welch observed of the Somerset & Dorset: 'This was unquestionably an absolutely splendid route which still seems to be virtually everyone's favourite more than forty years after its sad demise.'

The S&D was nicknamed both the Swift & Delightful and the Slow & Dirty, but nicknames of any kind are usually bestowed out of affection. The appeal of the S&D lies perhaps in the way it joined obscure dots: places with a commercial identity that globalisation has erased.

The Somerset & Dorset Railway was formed in 1862, when the Somerset Central Railway and the Dorset Central Railway amalgamated. The Somerset Central Railway arose because Clark's shoe factory at Street near Glastonbury wanted to export its shoes from Burnham, which was on the sea (or at least the Bristol Channel) but was not yet Burnham-on-Sea. After the amalgamation, the S&D line extended east to Poole, so it went diagonally from sea to sea. This opened up the following journey: people could travel by sea from Wales to Burnham; then they could take the train to Poole, from where they could sail to Cherbourg in France – all of which cut out the need to go around Land's End. Even so, there weren't many takers. Then the S&D directors had another idea: they could connect the North Somerset coalfield at Radstock and Midsomer Norton to Bath Green Park Station and northerly points beyond. So in 1872 they extended a tentacle north from Evercreech Junction to Bath. (North of Bath, this line intersected with the 'Bath branch' of the Midland Railway to Mangotsfield, which in turn intercepted the Birmingham–to–Bristol line. The Mangotsfield branch was closed to passengers in 1966, but some of the route is preserved by the Avon Valley Railway.)

The cost of building and blasting the line (seven viaducts and four tunnels) was ruinous, and in 1875 the Somerset & Dorset was leased to the Midland Railway and the London & South Western Railway, who operated it jointly. So the Somerset & Dorset Railway became the Somerset & Dorset *Joint* Railway. In *The Trains Now Departed*, Michael Williams evokes some of the countrified aspects of the line: 'A train would sometimes come to a halt in the middle of nowhere, much to the bemusement of passengers, while its crew picked mushrooms in a line-side field before frying them with bacon and eggs on a hot shovel in the firebox.' But he adds, 'Unusual among country railways, the S&D had its own dedicated express train.'

The Pines Express, which ran from Manchester to Bournemouth via Bath, was fascinating to railway enthusiasts. Most named trains started or ended in London. The Pines was the most famous of the cross-country ones, eclipsing, say, *The Devonian* (Bradford to Torquay and Paignton), or *The Cornishman* (Wolverhampton to Penzance). It had the romance of a holiday line, but *The Pines* was chiefly known for some freakish rolling stock. Michael Williams again:

> Few engines were powerful enough to tackle this mighty train single-handed, and some astonishing combinations were turned out ... Here was a modern Southern Railway West Country Class Pacific bunkered up to an elderly ex-Midland Railway 4F freight engine of Victorian provenance, or perhaps a tiny Edwardian 4-4-0 locomotive of a type which had mostly been consigned to museums. At the end of the 1950s there was a living cavalcade of British engineering history on view in all its glory for anyone who cared to spectate at the line-side on a summer Saturday morning. The locomotive department might

even throw into the mix the eleven distinctive 2–8–0 freight locomotives specially designed for the line by the Midland in 1914, which for years had trundled humble goods trains over the Mendip banks.

The S&DJ had become part of the Southern Railway in 1923. In 1948 it became one of the more exotic, and least profitable, items on BR's books. In 1962, *The Pines Express* was diverted over ex-GWR lines via Oxford, Basingstoke and Southampton. The old S&DJ had fallen into the hands of BR Western Region, and its fans suggest that the diversion of the *Pines* was an act of revenge by BR managers who still had a residual loyalty to the old GWR, which had always regarded the *Pines* as a trespasser on its territory.

We have already seen John Betjeman dancing about on the beach at Burnham in his film of 1963. He was part of a wider campaign to save the S&DJ routes, but in 1966 they fell to Beeching's axe.

It should be stressed that all of this drama took place on a territory whose most easterly boundary was 30 miles west of the West Somerset Railway, and of Washford station, where the Museum of S&D Trust is located. The Museum was closed, but my guide had admitted me to its main building. 'They couldn't acquire any of the trackbed of the S&D,' he said of the Trust, 'so they came to a Great Western line – ours, the West Somerset – that had closed in 1971. They were the first ones into this station, but they can't go anywhere; they're stuck.'

'It must have been heartbreaking for them,' I said, 'to see you re-creating your line, when they couldn't re-create theirs?'

He answered with a wry smile.

I suggested it was nice of the West Somerset people to paint Washford station green, in acknowledgement of the fact that the S&D lines had once been part of the Southern Railway (because I'd now figured out that this was why Washford was green).

This elicited only a modest shrug, and, 'Your average family from Butlins is not interested either way.'

There is a melancholia attaching even to this Southern green, since its introduction had meant the end of the S&D's famous livery, which was applied to both its carriages and engines: Prussian Blue. The Museum contained a beautifully restored carriage in that colour. 'A chap from the trust was playing cricket at Templecombe,' my guide explained, 'and he saw the carriage on the boundary. It was being used as a changing room.' It dates from 1886, and cost £57,000 to refurbish.

This is arguably the prize exhibit of the Trust; the other contender is S&D 7F class 2–8–0 number 88. This – one of the distinctive 2–8–0 freight locomotives mentioned by Michael Williams – sometimes runs on the WSR. The Trust also owns *Kilmersdon* – an 0–4–0 tank engine of the S&D. Otherwise the line is memorialised in well-presented fragments: a station lamp, newspaper cuttings ('Burnham Jetty Not Saved'). There are photographs, one captioned 'Our Volunteers at Work', showing the cleaning of a sign reading 'Museum'. There is a poem on the wall:

The services were cut right down
The Pines *diverted away*
'Til empty trains became the norm

And hastened that sad day.

The S&D Railway Trust is one of many organisations devoted to the S&D. There is the Somerset & Dorset Railway *Heritage* Trust (note the italicised word differentiating it from the organisation just described), which has a museum and a mile of track at Midsomer Norton. There is the Gartell Light Railway, at Yenston in the Blackmore Vale, Somerset ('run by the Gartell family and their friends'), which operates a 2-foot-gauge line along part of the trackbed of the old S&D. There is the Shillingstone Railway Project, which sounds like a 1970s prog-rock band and is based at Shillingstone Station, ex of the S&D. It is run by the North Dorset Railway Trust. Demonstration trains are planned. There is also the New Somerset & Dorset Railway, which aspires to re-open the whole S&D line from Bournemouth to Bath.

'God's Own Country'*

The North Yorkshire Moors Railway

The working assumption of Eric Tonks, that statistician of preservation, was that what attracted people to the preserved railways was the *railways*, but he made an exception of the North Yorkshire Moors Railway. The line 'possesses some of the finest scenery offered by a preservation system, and anyone can go again and again, just for the pleasure of the ride alone.' The NYMR has always been proud of its setting; its advertising slogan used to

*That's Yorkshire (at least as far as people from Yorkshire are concerned).

be 'Escape to the Moors.' I like the practical, railwayman's way its setting was evoked in an article for *Heritage Railway* magazine by Andrew Scott: 'Trains climb from sea level at Whitby to a 532-foot summit at Ellerbeck before descending more than 400 foot to Pickering. Locos have to work hard and there is no room for anything less than a Class 4 in regular service.* (Scott graduated in Civil Engineering from Newcastle University; he first became involved in the NYMR through that university's railway club, which has now gone the way of most university railway clubs. Today, having retired from running the National Railway Museum, he is Vice-Chair of the North York Moors Historical Railway Trust, and we will be hearing more from him in a moment.)

The NYMR was my local preserved line when I was growing up in York, and I often revisit it. I once did a five-day footplate experience course on the line. The instructor-fireman made his tea by pouring hot water onto tea leaves held in a twist of greaseproof paper, which nobody in the real world has done since about 1955. There was only one other pupil: a slightly melancholic chap, who'd had a bout of prostate cancer, which motivated him to get on with all the things he'd always wanted to do. We were referred to NYMR pamphlets about technical matters ('All new worsted trimmings must be thoroughly soaked in oil before use'), and these were called 'Mutual Improvement Guides', which is what they would have been called in 1900.

It was 2001, the year of the foot-and-mouth epidemic,

*That is, a BR Standard Class 4: a medium-sized engine used mainly for freight haulage.

and the countryside was off limits – to humans but not to sheep, and I enjoyed being at the regulator as they bounded away on either side of the engine, like a bow wave in reverse. In normal times, the NYMR is reputedly 'the busiest heritage railway in the world'. Its northerly station, Grosmont, where passengers often have to change for the onward trains to Whitby, can resemble rush-hour Waterloo, but in 2001 I more or less had the whole 18 miles of the line to myself. It was as if I'd gone back to an age when the NYMR line really was a sleepy branch, of the kind that had Dr Beeching reaching for his calculating machine.

I think of Ian Marchant, writing in *Parallel Lines*. He found the preserved lines' pretensions to capture the atmosphere of a real country railway undermined by the sheer weight of visitors. He writes, of a generic preserved railway:

> If the line had had this many visitors when it was opened, it would never have closed. And the railway enthusiasts are caught on the horns of this dilemma; they wish the line had stayed open, but then they wouldn't have been able to play trains; they are sorry that the line has closed, but they are glad that it has because they like playing trains.

Sometimes I think I prefer the NYMR on a rainy weekday afternoon in winter, when I might take the last-but-one train from Pickering to Grosmont. (Not the last one, because then I'd be stranded in Grosmont.) On those days, the countryside seems almost dangerous – lonelier, with steeper hills, the waters of the streams rushing faster

– and I might be the only person in a first-class compartment carriage. (Not sure I ever did achieve that, but I was once one of three in such a carriage, and the other two were asleep.)

But there's a festive aspect to the NYMR on a bright summer's day, and it seems completely logical that a friend of mine, the children's writer and illustrator, Caroline Golding, should – during every annual steam gala – do a round trip on the line accompanied by her partner and equipped with a fruit cake, sweets, a Thermos of tea, bone china teacups and half a bottle of champagne. 'Admittedly,' she once told me, 'the napkins are folded kitchen roll ... I *would* travel first-class,' she added, 'but I'm pretty sure even that won't get you a tablecloth.' When she's finished her tea, Caroline spends most of the rest of the journey waving at strangers through the window.

The presence of many children helps create this festive atmosphere, as on an old-fashioned seaside holiday. The railway likes this jollity, which seems a guarantee of its future. 'I won't mention any names,' Andrew Scott told me, 'but you go on some preserved lines, and you do wonder if anyone will still be there in ten years' time.' The NYMR goes out of its way to attract families, especially families of volunteers. Not only do you get three or four or whatever for the price of one but, says Andrew Scott, 'any junior volunteer scheme needs adults to supervise – and the parents can do that.'

The NYMR is mainly a steam railway, with some real thoroughbreds. It has Gresley's eponymous A4, *Sir Nigel Gresley*, close cousin of *Mallard*, and with a speed record of its own (fastest postwar speed: 112 mph.). Today, *Sir*

Nigel combines ambling along the NYMR with 75-mph work-outs along the main line. The line has some women among its volunteer footplate crews, one of whom – a thirty-two-year-old schoolteacher called Bess – I interviewed for a newspaper. 'I'm a fireman,' she said. 'There's no such thing as a fire*woman*. At the end of every turn, I have this big, inane grin on my face.'

I asked her to sum up the appeal.

'It's in the sway and rhythm of the engine. I listen to the beat. I play a bit of fiddle, and I think it's related to that.'

I also interviewed Andy, a twenty-two-year-old driver on the NYMR. (You can start training for the footplate at sixteen and be driving by the age of twenty-one.) I asked whether he'd ever been a trainspotter.

'That's one thing I'm not. I like the mechanical side. It's about power.'

(For the same article, I interviewed Henry, also twenty-two, who'd 'served his time' training as a steam fitter at the National Railway Museum, which seems an outrageous idea, as if modern-day Egyptologists had started building pyramids. He told me that some of the other men on the course were retired: 'In their lives they'd seen maybe sixty engines in steam at some depot in the 1960s. But I've seen sixteen in steam at one time, which isn't too bad.' Yet another Henry had done a mechanical engineering apprenticeship on the Bluebell Railway. He'd been raised in a cottage by that line, and his mother was a 'senior fireman' on the Bluebell and ran two of the three youth clubs at the Bluebell. This Henry's father, Michael, drove on too many preserved lines to mention. Henry himself was a volunteer fireman on the Bluebell, a driver on the North Norfolk and a fitter on the Kent & East Sussex.)

The NYMR carriages also make a nice turn-out, especially the ex-LNER 'teaks', and the line has real Pullmans for its dining trains. On the NYMR, everyone seems to get into the spirit of things. I once saw a driver leaning out of the cab of his tank, which was running light out of Pickering, while ruminatively smoking a pipe. (On closer inspection, he was vaping with a retro pipe-like device.) I also once saw a man with a bad leg on the platform at Pickering who walked with the aid of an antique wooden crutch. It was as if they were taking their cue from the fact that Pickering is styled as a 1930s station.

For the historical education of the punters all the stations on the line are fitted out to reflect different periods. 'Pickering', writes Peter Smith in *Station Colours*, 'is the perfect reproduction of an LNER station in the green livery – matt colours, just the right amount of weathering – superb! This is the one to copy.' The NYMR is the star of its own Channel 5 TV series, which depicts a series of contrived crises (will the crocked engine be fixed in time?) but is relaxing viewing nonetheless. The line has about 10,000 members, of whom about a thousand are regular volunteers – not as many as you'd think. Like most preserved lines, it is perpetually on the look-out for more. There are a hundred paid staff – which is a lot.

The NYMR arose from the economic problems of Whitby in the early nineteenth century. As the line's archivist once told me, 'It was all about fish.' Well, fish and whales (which are not fish). The whaling industry was in decline; Whitby wanted to sell more fish, and a railway connection to the nearest market town – Pickering – seemed the answer. This was advanced thinking, and the line (engineered by George Stephenson himself) was one of the first in Yorkshire, although for the first eleven years after its opening in 1836 it was worked by horses. In 1845, the 'Railway King' George Hudson's York & North Midland Railway Company took over the line and built a double track for engines. The York & North Midland also built a line from York to Scarborough, and this included a junction at Rillington, east of Malton, by which this line was connected to the Whitby–Pickering. (So a very scenic round trip was available: York to Whitby via Pickering, then down the coast from Whitby to Scarborough and back to York.)

By the late 1950s, the majority of Whitby–Pickering freight was going by road and the line had become mainly a tourist railway. Beeching envisaged the closure of all railway routes to Whitby, of which the town had accumulated three. After protests, Transport Minister Marples reprieved the Esk Valley Line, which connects two coastal towns, Whitby and Middlesbrough, although it is not a coastal line: it runs inland. But the lines from Pickering to Whitby and from Pickering to Malton were closed in 1966, leaving Pickering marooned.

In 1967 locals formed the North York Moors Railway Preservation Society and, with assistance from North Riding County Council, they bought the Whitby–Pickering line for £42,000. The first engine of the preservation era was cheered along the line by crowds one snowy day in 1969. Public services commenced in 1973, and in 2007 steam services were extended to Whitby, running over the Esk Valley Line.

Let us take a ride along the NYMR, imagining a sunny summer's day.

The line enters the National Park at New Bridge just north of Pickering. At first, the train is cosily enclosed in banks of purple heather. Assume we are riding in one of the 'teaks' – an open third. The interiors are in tawny shades. (In the late afternoon, you feel that the sunset has been trapped in these carriages, and the pinkish light is redoubled in the oval mirrors placed between the windows.)

We come to Levisham, where the station clock has metaphorically stopped at 1912. There's a tea room, an artist in residence, a camping coach – and this is the starting point for many footpaths into the steeply rising Moors. The platforms are connected by Levisham's own Bridge of Sighs. In *Britain's 100 Best Railway Stations,* Simon Jenkins writes: 'The station is overlooked by a delightful rust-coloured footbridge. Its soft curve, elegant with lattice panels, is a model Victorian work, a lesson to all who feel modern bridges should be made of straight lines and boxes.'

It is with a note of pride that the guide to the line reports that the actual village of Levisham is a mile away. As for the next station, Newtondale, that's 'further from a public road than any other station in England'. There's a green and white wooden shelter. A sign reading 'Way Out' points to a thicket of trees. After Newtondale the view opens up and the train begins to climb into a nature reserve 'with a unique mixture of plants, dragonflies and many bird species'.

We are approaching Goathland and the year 1922

(the station's theme). The village of Goathland doubled as Aidensfield in Yorkshire TV's *Heartbeat* (sort of *All Creatures Great and Small* but with criminals). Younger generations will recognise it as Hogsmeade from the first Harry Potter film. It seems to me that the NYMR is often in films, but then I'm looking out for it, and I recognised the supposed 'King's Cross' at the start of the *Downton Abbey* film for what it really was: Pickering Station smothered in CGI. But Andrew Scott told me the income from film and TV was 'patchy'.

Surely, I suggested, it's even patchier everywhere else?

'Not at the Bluebell,' he said. 'You can hardly turn on a *Poirot* or a *Miss Marple* without seeing that. But then it's much nearer to London.'

On my trips to the Moors, revisiting boyhood scenes, I often stay in Goathland, at the Mallyan Spout Hotel, which is like staying in the country house of a particularly easy-going friend. Every evening, I walk down the hill to the nearby hamlet of Beck Hole for a drink at the Birch Hall Inn. Every time I look at the *Good Pub Guide* or similar, the Birch Hall seems to have a place of honour at the very beginning, cited as the ideal towards which all the other entries are aspiring. There are two tiny drinking rooms; one has wallpaper – slightly sooty near the fireplace – showing huntsmen on horses. Leaving aside drinkers in the pub, I have never seen any people in Beck Hole, only sheep, but until 1951 the village was served by a freight branch off the Whitby–Pickering, because there was – and is – ironstone hereabouts.

We come to Grosmont Station, which is decorated as a BR North Eastern Region station (the North Eastern being the predecessor of the Eastern Region, which was created in 1967). The station colours of the North Eastern Region were Oriental Blue and ivory. In *Station Colours*, Peter Smith includes pictures of Grosmont Station, observing, 'As far as I'm concerned this is the best, most imaginative colour scheme ever applied to a British railway building.'

At Grosmont, a pleasant hiatus occurs. Those passengers terminating here have a look around before going back down the line, or they might have to change trains if they want to go on to Whitby (some trains do go direct); or people might head off into the moors. An old wooden sign just south of the station has been branded with a primitive map, along with 'Picnic Ground', Look at Wildlife', 'Beautiful Woodland', 'Riverside Paths'.

There's the narrow tunnel through which the old horse-drawn line used to run, and at the end of it today is the locomotive shed, of which you can sometimes have a tour. Grosmont used to be *called* Tunnel, a perfunctory name, but at the time it was not a tourist destination, and it continued not to be one for much of the twentieth century. The presence of ironstone in the surrounding countryside made it a place hard to imagine but very characteristic of Yorkshire: an industrial village, overshadowed by a giant blast furnace that would have looked more at home in Middlesbrough.

Insofar as the real world impinges on Grosmont today, it's represented by the Network Rail station, which abuts the NYMR one. Both are called Grosmont. The Network Rail one serves the Esk Valley Line, so you can use it for Whitby or Middlesbrough. The NYMR's station will take you back to Pickering, or on to Whitby via the Esk Valley Line in summer. Not all the NYMR's steam engines are suitable for the Whitby run. 'They have to have all the gizmos to deal with modern signalling,' an NYMR driver once told me.

It's 10 miles from Grosmont to Whitby, the line crossing and re-crossing the river Esk. Whitby Station, designed by George Andrews in 1854, reminds me of Santa Lucia in Venice in that it is right on the water. There's a photograph by Francis Meadow Sutcliffe, the chronicler of Victorian Whitby, showing a tall sailing ship framed by one of the station porticos: a conjunction of old and new transport.

My most recent journey back from Whitby to Grosmont was not by steam train but by Northern Rail in a packed Class 156 Super Sprinter. As if to compensate for the modest accommodation, the guard delivered all his announcements in rhyming couplets. On leaving Whitby, I heard something like:

Safety information posters contain useful data
So please read the instructions – I'll ask questions later.

The speaker was a Northern train guard called Graham Palmer who attempts – successfully, I think – to lighten the bureaucratic strictures of modern train operation with a bit of poetry.

At the end of that day, I was walking through the old-line tunnel – pleasingly cool and dark on a day of blazing heat – towards the engine shed. I fell into step with a young man who had been driving or firing. He was a big chap, soot-stained, with a steady, lumbering walk. I got talking to him, telling him about this book.

'Have you tried the Wensleydale?' he asked. 'I drive there on the odd occasion. I sometimes think it's even better than here.'

'Why?' I asked.

'More scenery. We're in a valley, they're not; and their line is longer.'

When I got back to Pickering, I walked away from the station in a southerly direction. This was the direction of my parked car, also of Malton, and the Rillington Junction that had once connected Pickering to the wider world. Will the NYMR reconnect to Malton? In *The Railway Preservation Revolution,* Jonathan Brown mentions that some local authorities 'have wanted to move more quickly than the railways themselves: councils, chambers of commerce and others have been actively pressing for the North Yorkshire Moors Railway to extend southwards from Pickering to Malton, something which has not been at the top of the railway's priorities'. One volunteer told me, 'We'd be overwhelmed; the whole character of the line would be changed.' Back in the 1960s Pickering Council wanted to demolish the station. Today, the NYMR has been so successful that the possibility of too *much* success can be envisaged.

The Wensleydale Railway

The Wensleydale Railway is a sleeping giant among preserved lines. I think of it as being like Scarborough: tremendous natural assets but scruffy and undersold. It runs – almost – between two more famous lines: the East Coast Main Line to the east, and the Settle–Carlisle line, which is not a 'preserved line' even though it was consciously preserved, and a high percentage of its passengers are riding on it for pure pleasure. The branch line on which the Wensleydale is based did (from 1878) connect those two lines. It carried limestone, milk and not many people, and was closed piecemeal between the mid-1950s and the early 1980s. By the early 1990s, all that remained were occasional trains running between a quarry at Redmire and Northallerton.

The Wensleydale Railway Association was formed in 1990. The track west of Redmire had been lifted, but the Ministry of Defence paid for the refurbishment of the rest of the line, which it wanted to use to transport military vehicles – via Northallerton – to Catterick, where there is a large army camp. These trains continue to run, which is why the modern-day Wensleydale Railway is marked in the excellent *Railway Atlas Then and Now* as both pink (for a preserved line) and red (for a freight-only line). In 2000 the WRA formed the Wensleydale Railway Plc in order to make a share offer, by which £1.2m was raised. The line between Redmire and Northallerton was leased from BR, and passenger services resumed in 2003. In its present form the line is 22 miles, which is *long*. It doesn't quite reach Northallerton but stops a fifteen-minute walk away at a new, temporary station called Northallerton West, although at the time of writing WR trains are not even reaching Northallerton West, following an accident at a level crossing. Pending the upgrading of signalling at that crossing, the current easterly terminus is at Leeming Bar. The WR aims to restore the track west of Redmire, so re-connecting to the Settle & Carlisle line, which will be effected by reaching the station formerly called Hawes Junction* and now called Garsdale.

At Leeming Bar, the long platform was deserted; some

* Where a terrible 'smash' once occurred – see 'Christmas Day'.

BR Mark 1s and a few diesels hung about in grassy sidings. The centre of operations was a BR maroon carriage in a bay siding; there was one man inside, selling tickets. I asked if any train was due.

'Yes,' he said, 'but I'm afraid we only have a Class 37 running today.' He obviously assumed that I required steam traction in return for my cash, but I find the Class 37s quite glam: like Deltics, they have a bonnet or a nose, which gives them a purposeful, American look.

I spent an enjoyable afternoon going to Redmire and back, with the Class 37 living up to its nickname, which is to say growling (they're known as 'Growlers'), and also sounding – given the rural context – like a tractor, which is the alternative nickname for Class 37s.

The WR averages a creditable 50,000 visitors a year, but on the day of my visit it was very quiet, therefore unlike the NYMR. The scenery was also different – rangier views, as promised by the NYMR train driver. This was the Dales, and a dale is a broad valley. There were frequent outcroppings of limestone on the hills, and the limestone seemed to be going back into the earth in the case of many crumbling barns and drystone walls. There were only half a dozen people on the train, most of them volunteers. They were talking about busier days. ' ... Two hundred people we had on the train for that lady's seventieth ... We had canapes on the train and a hog roast ... We've got the prosecco train in July; all the young girlies come on that.' The conversation then turned to the question of whether one elderly female volunteer ought to continue her volunteer role as crossing-keeper, and I recalled reading of a stipulation in the Royal Train operating instructions, as described by Brian Hollingsworth in *The Pleasures of*

Railways: 'Even as recently as fifteen years ago [he was writing in 1983] one found in the Royal Train instructions such phrases as: "At level crossings where a gate-woman is in charge, a competent man shall be present half an hour before the Royal Train is due."'

I got off the train at Leyburn Station, whose very substantial platform buildings resembled one side of a quiet street, or a terrace of houses. It was now late afternoon. The man who ran the shop was just packing up; the café had already closed. Two tired backpackers waited with me for the train to come back from Redmire. I read a poster:

Vacancies for Volunteers

Trackside maintenance
Painters
Crossing keepers
Gardeners
Decorators
Guards
Office staff
Train drivers
Cleaners
Engineers

Volunteer and Start Your Life All Over Again

The one that caught my eye was 'train drivers'. I find the idea of being in charge of a steam engine daunting, but I think I could handle a diesel, and the WR has a dozen of those, as against a handful of steam engines. I fantasised for a while about retiring to the Dales and the Wensleydale Railway. It seems to me that the WR is about to go places. It has just received a grant to redevelop Leyburn station, and when the proposed reconnections are made at either end, patronage will presumably soar. Here is a chance to go back into the past with real momentum.

The Embsay & Bolton Abbey Steam Railway

The Embsay & Bolton Abbey Railway attracts 100,000 visitors a year, but on the day of my visit it seemed as pleasantly sleepy as the Wensleydale. Before setting off from London, I'd phoned the railway, and had quite a long chat with its commercial manager, Peter Walker. He told me I could get a good flavour of the line pre-preservation in a memoir called *Bolton Abbey: The Time of My Life*, by Donald Wood, whose father was the blacksmith at Bolton Abbey. Here's Donald Wood's recollection of travelling on the 8.10 train from Bolton Abbey to Skipton, where he was a pupil at Ermysted's Grammar School. The boys for the grammar school would travel in separate compartments from the girls travelling to Skipton High School ...

> Once on the train, of course there were high jinks. Little ones were sometimes lifted onto the net luggage rack, and we often took the sticks out of the roller blind to fence with. On one occasion Robert Wilson went out of the door of the carriage when the train was going through Haw Bank, Embsay, tunnel, walked along the running board and got into the girls' carriage.

Mr Walker then spoke with bemusement about an

author of historical novels who'd set an early-twentieth-century scene at Holywell Halt station, which was not on the line at that point. 'We built that station, and we opened it in 1987,' he said.

Mr Walker said he would leave a complimentary ticket for me at Bolton Abbey station, but having forgotten that geographical specification, I turned up at the other end of the line – Embsay – expecting to find the ticket. Naturally, the young ticket clerk couldn't find it and, being new, he'd never heard of Mr Walker. 'It doesn't matter, though,' he said, and he gave me a free ticket anyway. The Embsay & Bolton Abbey bills itself as 'the friendly railway', and with some justification, I would say. Its other boast – 'An atmosphere of the days of the country branch line prevails' – is also justified.

Embsay Station is done in one of the many and confusing station liveries of the London Midland & Scottish Railway: cream and green. A train awaited with some worn-out maroon BR Mark 1s behind an ex-industrial tank engine of some sort. The railway has a lot of those. I got talking to three very laid-back engine drivers, all drinking tea on a platform bench. One of them said of the industrial locos, 'We've got 'em because they're cheap. We're not really an industrial *railway*.' Very close to the railway, however, and running north of it, is what must be the most beautiful of the modern-day industrial lines. It runs up Wharfedale to a limestone quarry at Swinden near Kilnsey, a village overlooked by a huge limestone cliff. This line is called the Yorkshire Dales Railway, which is what the E&BA was called when it was first reopened back in 1981.

The E&BA runs over 4 miles of the line from Ilkley

to Skipton, which closed in 1965. The E&BA aims to reconnect with Skipton, in whose Network Rail station a disused platform awaits it, and this would involve renewing the junction with the above-mentioned freight line, which the E&BA would also rather like to take over, should it ever cease to carry the limestone.

I walked into the station shop at Embsay, which carries a range of bric-a-brac and books. A satisfied customer was saying to the assistant: 'You've some very interesting items. We've never not bought anything here.' The assistant nodded graciously while carrying on with her knitting.

I rode the train through slumbering fields to Bolton Abbey, where a young woman came out of the ticket office. 'Are you Andrew Martin?' she said. 'We've got a ticket for you. We wondered where you'd got to.'

Sidings near the station were full of mouldering diesels – the scene was quite apocalyptic, and I was reminded of something I'd read about accounting practices on preserved railways. It was difficult to define their assets. Did a crumbling engine that would cost a fortune to repair count as one?

In the 1920s, Bolton Abbey Station might have been packed on a summer Saturday like this, with well-heeled day trippers arriving to go on to the Abbey a couple of miles away. According to Donald Wood, in *Bolton Abbey: The Time of My Life*, local urchins would crawl underneath the planks of the platform to wait for coins to fall between the cracks. In the station yard, Wood recalls, 'there were about 14 wagonettes, driven by local farmers' to take people to the Abbey. An old guard's van was their cab-man's shelter. Later in the 1920s, the farmers began to offer rides in Model T Fords, charabancs or motorcycle sidecars, but at the same time, people were bypassing the railway altogether and taking buses to the Abbey from Skipton and Ilkley.

The route to the Abbey was explained on a blackboard at the station. Today, there are no cars waiting in the station yard to take you there, but there are all too many screaming up and down the A59, and the noise they make is the soundtrack for the first stage of the walk to the Abbey.

The Strathspey Railway

Problems in England

Scotland has only half a dozen preserved railways, most of them small. In *The Railway Preservation Revolution*, Jonathan Brown puts this down to 'sparse population in rural areas' and 'declining industries in many other districts'. Or we might say that the railway romance catered to in England by the preserved lines is met in Scotland by some surviving national-network lines: the Kyle Line (Inverness to Kyle of Lochalsh) and the West Highland Line (Glasgow to Mallaig and Oban), which daydream their way through the remote West Highlands, and the Far North Line, which makes its lonely way over the vast marsh that is the Flow Country. Beeching wanted to excise all of them, of course, along with most Scottish branch lines, but he often didn't get his way in Scotland.

One late autumn, I rented a cottage near the Far North Line in order to finish a novel. I forget the name of the nearest station: it was about the twentieth one north of Inverness, possibly Kinbrace, and part of the line was visible from the cottage, which was otherwise completely isolated. I rented it from the owner of a big estate; I never saw him during my stay, only his gamekeeper, who arrived on my first day with some sacks of coal, and the coal fires I burned helped make up for the disappointment of there being only two-car diesel Sprinters on the Far North Line rather than steam engines of the Highland Railway, or its successor on the route, the LMS. One day, I cycled to the line and took a train to Thurso, which was smothered in fog, with Orkney looming vaguely but massively from the sea.

The principal preserved railways in Scotland are the Bo'ness & Kinneil on the Firth of Forth, and the Strathspey at Aviemore. Given that I am writing about railways in the landscape, I thought a Highland line should be included, so I boarded an LNER train – and, given the retrospective nature of this book, I should make clear that I mean the new LNER, the London North Eastern Railway, as opposed to the old one: the London & North Eastern Railway. The name of the new company is sufficiently similar to that of the old one to impede any internet search for the old one, but the new one lacks the conjunction, which you do need. Without it, as Ian Jack wrote in the *Guardian*, the suggestion is of 'a suburban transport network that reaches no further than Enfield'.

The 10 a.m. LNER train to Edinburgh was ten minutes late out of King's Cross; it was packed, probably because of the Edinburgh Festival. This was August, after all – and yet heavy rain was falling. I had to stand, which was uncomfortable because I had a bad back. The woman alongside me, who was sitting down, seemed to express my thoughts as she talked to her friend: 'It's so miserable isn't it? I mean, *we* have seats so we're lucky in a way, but the seats are tiny, and it takes so *long*. It's so tempting to fly and it's half the price you know, which is ridiculous, given the environmental impact. Quite honestly, you know [with a glance at me] I would *refuse* to stand. I think you can demand a seat in First Class. You can say, "I have a condition, and I *have* to sit down."' One thought of mine she didn't express: this service – the 10 a.m. from King's Cross to Edinburgh – was once *The Flying Scotsman.*

At York I got a seat, but by now my pleasurable anticipation of the Strathspey Railway had darkened. 'It had better be good,' I thought. We were forty-five minutes late into Edinburgh Waverley, so I missed my connection for Perth. I had assumed I had to change at Perth for Aviemore, and this indeed is what the little timetable I picked up in the ticket office at Waverley said. But this timetable was wrong. I could get to Aviemore direct by a train departing in an hour and half's time, but that train – another LNER service – was running an hour late.

'It's down to problems in England,' the Scottish ticket clerk said.

Somewhere around Perth, I did get a seat on the train to Aviemore, which was as crowded as the one from King's Cross. The passing scenery seemed to have a morbid grandeur. A persistent thin rain was falling. You wouldn't call it drizzle; it was more malevolent than that. The clouds, somersaulting about amid the hills, were ceasing to observe the distinction between earth and sky. I kept seeing broken wooden bridges on the adjacent river. On the adjacent road, a car transporter had pulled over, obviously in distress. One of the cars on its top deck was dangling dangerously.

Unlike, say, the North Yorkshire Moors Railway, the

line from which the Strathspey Railway originates was located in a dramatic landscape *because* of that landscape; it wasn't just incidental to the railway. Aviemore Station, first opened in 1867, rebuilt in 1898, is in the cinnamon-and-white, wooden-chalet style of the Highland Railway: it was meant to appear as a pretty and exotic holiday destination to the tweedy types turning up with their fishing rods and guns. In *Britain's 100 Best Railway Stations*, Simon Jenkins writes that 'it might be a set for a Hollywood winter sports musical ... Cottagey facades and bright colours hint at Zermatt or Val d'Isère.' LNER and Scotrail share the station with the Strathspey Railway, which has Platform 3.

Aviemore – as modernised for tourism in the 1970s – is a disorientating place, with something of Los Angeles about it, but with the Cairngorms to the east instead of a desert. Opposite the station is the turreted, baronial Cairngorm Hotel, which is the kind of thing you expect to find, but it looks like a pastiche when you factor in the massive Tesco's, the shopping mall, a very obtrusive Shell filling station, an Italian ice cream shop, Mambo's Café Bar, and a sense (at this time of night) of post-après-ski lassitude. In the mid-sixties a brutalist Activity Centre – now the Macdonald Aviemore Resort – was built on the site of another baronial hotel. I don't like the word 'activity': it suggests school trips, things undertaken against one's will.

The Monarch of the Glen

The main activity at Aviemore is skiing; there's also quad-biking, mountain biking, hiking and rock climbing. The people sitting near me at breakfast in the guest house might have been doing any of those except skiing, since there was no snow on the Cairngorm peaks: their talk was of things like muscle tone, stress fractures and the necessity (up to a point) of carbohydrates. As I approached the Strathspey Railway in the still-falling rain, I was hoping it would take me away from this modern, strenuous Aviemore, and back to the old one, which I imagined was a gentler place (unless you happened to be a grouse or a trout) with romantic, John Buchan-esque overtones.

I was early for the first train, but the waiting room on Platform 3 was promising, with a disparate collection of old armchairs – like the cluttered sitting room of a farmhouse. I sat down and read about the history of the line. The first route between Perth and Inverness opened in 1863, going via Aviemore and (north of Aviemore) the oddly named Boat of Garten. In 1897 a more direct route was opened, going in a more westerly direction via Carrbridge. Aviemore was fortuitously placed at the junction of the two lines, and this was the start of its popularity, and the cue for the expansion of the station. The 1863 route was closed under Beeching in 1965. What is today the Strathspey Railway runs along a 15-mile stretch of that 1863 route – between Aviemore and Broomhill via Boat of Garten. Let's just say for now that it opened in 1978, but its roots are entangled with the preservation group set up in the 1950s, and I will say more about this when discussing freight and industrial railways.

The train came in: a collection of BR Mark 1s and 2s with seating not necessarily original, hauled by an Ivatt Class 2–6–0 built in 1952 by BR to an LMS design. The driver was twenty-one years old and obviously had 'the right stuff' (competent, phlegmatic). He'd been a volunteer on the line since he was thirteen, and now he was a paid driver. I asked if he'd always wanted to drive steam engines, and he shrugged: 'I didn't really have a plan.'

Did he mean to make a long-term career of it?

'We'll see.'

Did he enjoy it?

'Sometimes. This is probably the best engine to drive in bad weather.'

I asked why.

'Because it has a full cab.' (He had a roof over his head, in other words.)

I wondered what the worst engine was for bad weather.

'828,' he said. 'It was built by the Caledonian Railway in 1899.'

I asked if he was sometimes driving in falling snow.

'Yeah.'

I suggested that he was lucky to be paid to drive, and he nodded, adding, 'We don't really have enough volunteers on the line.' I wondered how many active volunteers (as opposed to paid staff) the line had.

'About ten.'

And how many would the line have ideally?

'About a hundred,' he said, and laughed.

The carriages were almost all with open seating as against compartments.

'Open's better for the scenery,' the platform guard had told me, adding, as he pointed to the west, 'Most of it's on that side.' He was indicating the Cairngorms, which were becoming more delineated as the rain cleared. As the train pulled away, I walked along to the brake van – which requires a word on terminology. On a freight train a 'brake van' or 'guard's van' is a separate vehicle at the end of a rake of carriages. On a passenger train what's generally called a brake or guard's van will be a space within a carriage containing a hand brake and often parcels and bikes, and a guard.

On this train, I became acquainted with the guard and the steward: both well-preserved, intelligent men in their early seventies or so. The steward was paid ('But don't tell anyone!'); the guard, who had worked in IT until retirement, was a volunteer. I asked if he would be doing it if the countryside weren't so attractive.

'Yes, I would. It passes the time, and I like meeting people.' He was interested in the countryside nonetheless. He explained that a strath is a wide valley, so we were in the strath of the river Spey. Boat of Garten? That meant there had once been a river crossing at that point. 'But nobody knows what Garten means.'

Beyond the river lay purple heather, then woodland, then the blue and grey Cairngorms, where about four different types of weather were occurring simultaneously. It was still raining on the highest peaks, I think. The guard indicated the gap to the left of Cairngorm: the Larghugru Pass: 'It means pass of the cattle.' I thought I could make out the rain being driven through it with great force. 'You've got a-hundred-and-twenty-mile-an-hour winds there; wind chill of minus twenty. When I walked through it one summer, I realised that my supposedly wind- and rain-proof clothes were nothing of the sort. If you walked through it in that,' he said, indicating my light rain mac, 'you'd die'; and he chuckled.

We were passing Boat of Garten golf course. It was undulating and prettily dappled by sunlight filtered through Scots pines and birch trees. If playing there, I would lose my ball immediately. 'The engine driver tosses the odd ball onto the fairway,' said the guard, 'just to confuse the issue.' He was emerging as a man with a pleasingly dark sense of humour. The steward told me that they often have people on the train with bikes, and sometimes fishing rods, but never with guns – the shooting is an entirely private business. At Boat of Garten, we collected a coach party – a common occurrence, which meant work serving coffee for the steward.

After Boat of Garten, the land near the line becomes agricultural. Russet-coloured cows watched us complacently. The steward returned after we'd set off again, to indicate the abandoned trackbed of the Great North of Scotland Railway (a small concern, the name notwithstanding), which had served distilleries to the east. The whisky was brought to Aviemore for transhipment to the south, 'and quite often a barrel of it would go missing.' We came to Broomhill, some of whose station signs also declare it to be Glenbogle, a fictional place invented by the makers of the TV serial, *Monarch of the Glen*. The Strathspey Railway Charitable Trust is building a 3½-mile extension beyond Broomhill to Grantown-on-Spey.

I was sorry to return to Aviemore: the Strathspey Railway *had* made me feel like a character in a Buchan novel. I had felt I ought to be proceeding to a rendezvous with a mysterious Highland laird, all Caledonian spruceness and amiability on the face of it, but with a dark secret to be uncovered.

By the Sea

The North Norfolk Railway

The old Midland & Great Northern Railway sprawled across an otherwise fairly empty part of the map of Britain, stretching from east Lincolnshire into north Norfolk. That this was an eccentric, picturesque network is suggested by the fact that John Betjeman was joint president of the society that revived part of it as the North Norfolk Railway in 1975. I include the North Norfolk as a seaside line because its climax is a spectacular terminal station in the middle of the seaside town of Sheringham. At the other end of the 5-mile line is Holt, with stations in between at Weybourne and Kelling Heath Park.

Millions of people have seen Weybourne Station without realising it. The *Dad's Army* episode, 'The Royal Train', was filmed there in 1973 during the process of the station's revival, Weybourne masquerading as Walmington-on-Sea. That's the one where Mainwaring's men are splashed by an engine supposedly taking on water from a water trough. It's funny, but anomalous: *Dad's Army* was set on the south coast, so the local railway would have been the Southern Railway, which never had any water troughs, since its trains never had to travel far enough to justify them. Nonetheless, Norfolk is *Dad's Army* territory. The exteriors of Walmington were filmed at Thetford, which has a Dad's Army Museum, and is surrounded by the landscape you see in the closing credits: 'the Brecks', with pale grassland, sandy soils and conifers. You see a similar landscape from the North Norfolk Railway.

At Holt Station on a misty morning, a sign informed me that the first train to depart would be 'STEAM hauled' by an LNER B12, 8572. The mist was pierced by the lights from an outbuilding in which the model railway display was housed, and by those from the building housing the Midland & Great Northern Joint Railway Museum. A booking-office clerk had just lit the fire in the booking office. Holt Station, prettily enclosed by trees of a Breckish type, is very well kept, almost manicured. It is hard to believe it is not the original Holt station, but that was demolished, and this preserved one was assembled piecemeal from buildings brought in from elsewhere. The station is liveried in the green and cream of the LNER, which took over the MGNJ network in 1923. It did so along with the Midland Railway, so the MGNJ remained 'Joint' until late in the day – until 1936, when the LNER took over full control. I do have to break it to the North Norfolk people that in *Station Colours* Peter Smith writes of Holt, 'The colours are right, but the modern gloss paint looks too bright and shiny, nothing like the old lead-based paint that would have appeared.' There is this consolation, though: 'To be fair, it was newly painted, so in a year or two it may have toned down a little and look OK.' He wrote that in 2016. Holt Station looked OK to me in 2018.

While waiting for the train, I went into the museum, which the curator opened early especially for me. The history of the 186 miles of the MGNJ is very intricate. I offer this summary from the *North Norfolk Railway Souvenir Guide:*

> The M&GN began as a series of locally promoted lines that were built both to serve impoverished rural areas and to challenge the monopoly of the Great Eastern Railway in West and North Norfolk. Mergers over

> the years created a sprawling cross-country network which was unified in 1893 by the formation of a Joint Management Committee ... of the Midland Railway and the Great Northern Railway.

In spite of being extensive (186 miles), the railway troubled few large towns, and most of its lines were single track. The works were at the incredibly rural-sounding Burgh Parva, just outside Melton Constable – which itself was a mere village, but also an important railway junction whose station had a platform 800 feet long. Today Melton Constable is marooned amid vast fields, with no railway. The *Souvenir Guide* notes, 'Melton Constable today with its railway terraced streets is more reminiscent of the railway towns of the industrial north of England than a small village in deepest rural Norfolk.'

We can't blame Beeching for the closure of the MGNJ: almost all of it was closed by BR as early as 1959. But we *can* blame him for the closure – in 1965 – of the stretch between Melton Constable and Sheringham that had survived the 1959 massacre and is now, in part, the NNR. That stretch had endured on the ragged remains of the holiday business that had once boomed, owing to the popularity of North Norfolk – dubbed 'Poppyland' by a Victorian journalist – with the more genteel type of tripper.

The preservationists had originally wanted to preserve a large part of the old MGNJ, but when the Poppyland stretch gradually became available they focused on that. In 1969, the newly created North Norfolk Railway Company became the first preserved railway to offer shares, raising £14,000. After years of toil on the overgrown trackbed, the line between Holt and Sheringham was opened to the general public in 1989.

Incidentally, the only part of the MGNJ that survives on the national network runs from Sheringham to Cromer. This is now the most northerly part of the Bittern Line to Norwich. I had spent part of the evening before my arrival at Holt in Sheringham, eating fish and chips at a restaurant on the main street as I watched rain lashing the town. I then walked to the front, to watch the great, grey waves slamming against the concrete abutments by which Sheringham is protected from erosion. I walked back along the main street to the two Sheringham stations. On one side of the road was the station of the preserved NNR: a pretty, green pavilion with a wide valanced canopy. Gold Flake cigarettes were advertised in one of those fatally seductive posters; a 'Ladies' Waiting Room' was indicated on a blue enamel sign. On the other side of the road was the modern-day Sheringham Station, operated by Greater Anglia. The station is just a platform, with a couple of pale blue lampposts and a 'Help Point'. The NNR station dates from 1887; the Greater Anglia one from 1967.

So here is a stand-off between old and new. It's not quite Ian Marchant's duality of the 'real' railway and the railway of dreams, because the Bittern Line – a community rail line – is quite dreamy, even if the infrastructure at Sheringham is not, and there is a rapprochement between the two, since in 2010 the level crossing on the main street was reinstated by volunteers and contractors, funded by donations from enthusiasts and local government. Occasional charter trains do run onto the NNR from the Bittern Line, and vice versa. It was in order to close and get rid of this level crossing that BR created a new station at Sheringham when the old one was closed ... so the whole thing is very British, what with the short-sighted transport planning, the persistence of the past, the determination of the volunteers.

I rode the line from Holt to Sheringham in a maroon BR Mark 1 coach of a special type built in the 1950s to serve London commuters. The carriage was a composite, meaning it accommodated both first- and third-class passengers. 'First' people could have no anxiety about an invasion from Third, since it was not possible to walk from one side of the carriage to the other. Each side had its own compartments and a short, wood-panelled corridor leading to a lavatory. The lavatory for the first-class side was so enormous that I almost expected to see an en-suite bath. I assumed (indeed hoped) that the lavatory in 'Third' was less impressive. In my compartment, the seat moquette was a silvery blue with gold horse chestnut leaves – one of the flamboyant 'Festival of Britain' moquette designs created by BR in 1951. Above the seats were maps showing the lines out of King's Cross, which only served to emphasise that London was a very long way off.

The sea comes into view very shortly after departure from Holt: you have the wide fields, then the wider sea, giving a great sense of freedom. We came to Weybourne Station, from where Weybourne itself is a speck in the distance, denoted by a church spire and windmill. I had been staying in Weybourne, and the previous day had set off to walk to the village station but gave up when I realised the distance involved. My accommodation was a rambling, spooky Georgian hotel in which I was apparently the only guest. The *Dad's Army* cast had stayed there when filming 'The Royal Train'. Late at night, as I negotiated the labyrinthine creaking corridors to my room, I formed the idea that they might still be in residence, albeit posthumously.

Weybourne Station is painted in the station colours of the old MGNJ: pinkish cream (which the railway inaccurately called 'stone') and yellowish brown, which it called 'tan', and which I think was not quite the same colour as the engines of the line, which were painted 'Autumn Leaf' or 'Golden Ochre'. Weybourne gets the thumbs-up from Peter Smith as 'the best place to see' the station colours of the MGNJ.

On arrival at Sheringham, I spoke to the fireman of the B12. He was called Adam and he was thirty-four. His grandfather had been a steam fitter, hence his interest. Adam was also mechanically minded and worked in car maintenance. I asked which he preferred: cars or steam engines.

'This,' he said, indicating the B12: 'a hundred per cent.' He can drive as well as fire, and he'd thought about going full-time as a steam driver. 'You can do it on the North York Moors or the South Devon, but you can't buy a house. You're on fifteen grand a year, whereas I can get thirty from fixing cars.' He told me his mate the driver 'was full-time on this railway, but this is his day off,' a statement that confused me. It turned out the driver was a full-time engineer with the North Norfolk, based at its engine shed at Weybourne. It was his day off from that, hence his appearance on the footplate of the B12. This says a lot about the esprit de corps on the North Norfolk Railway.

The Isle of Wight Steam Railway

When considering preserved lines by the sea, the Isle of Wight Steam Railway was uppermost in my mind. It wasn't until I was bounding across a very turbulent Solent in the catamaran, with freezing rain flying at the windows, that a doubt crept in. The doubt wasn't weather-related; it was brought on by having glanced down at the photocopied pages of my *British Railway Atlas Then and Now*. The Isle of Wight Steam Railway is about as inland as you can be on an island measuring 23 miles by 13.

The 'Then' map, of course, was the more crowded. The Isle of Wight demonstrated, in concentrated form, the virtues and the silliness of the Victorian free market in railways. By 1900 there were 55 miles of railways on the island, built by a dozen different companies, most targeting the tourist trade. I had in my files a summary of these railways, produced by the Isle of Wight Steam Railway. Some are despatched pretty quickly: 'It is difficult to say anything complimentary about the Isle of Wight (Newport Junction) Railway, which ran from Newport to Sandown. Several times the authorities refused permission for opening because of poor construction standards, and soon after completion the company went bankrupt.'

Lines on the island began to be closed from 1952. There were further closures in 1966, until only the line from Ryde to Shanklin remained, the islanders having fought hard to retain it. The line was then briefly closed for electrification. In the process the track was raised (to avoid flooding), reducing the clearance in the Ryde Tunnel by which the trains escape the town to head south. The small bore of that tunnel means that only ex-London Underground trains will fit, so these operate on the line, and one was waiting at the Ryde pierhead as the catamaran docked. As I walked towards it, with the sea bucking about amid the pier girders, I realised that this – the Island Line – was the service I had been vaguely thinking of when I booked my ticket from Waterloo to Portsmouth and looked up the catamaran timings. Well, the Island Line is a preserved line in a sense, so I will say something about it.

As a boy, I came on family holidays to the Isle of Wight. I liked to think of this as going abroad, but instead I found a sort of playground of skewed and whimsical Englishness. In York we had red buses, as in London. On the Isle of Wight they had green buses, strangely and rather intellectually named Southern Vectis. We always used to stay in Shanklin, which we reached by train. (There were no trains to Cowes or Ventnor, so we never went there, simple as that.)

In Shanklin, there were thatched cottages that seemed overwhelmed by their thatch, like the cottages on model railways ... and there were the ex-Tube trains of the Island Line, which in those days wasn't called the Island Line; it was just part of BR. I had been on Tube trains in London, so I knew they didn't belong on the Isle of Wight. I remember boarding them barefoot, and there'd be sand on the maple wood floors. Many years later, I had a flashback to those incongruous moments when I saw a man carrying a surfboard down an escalator at Piccadilly Circus.

The Island Line trains are 1938 Tube Stock, so they're sixty years out of date. It's the kind of time lag they like on the island: the previous Tube trains used on the island were also about sixty years out of date when they were replaced in 1992. Those were of Standard Tube Stock, which had been manufactured between the mid-1920s and 1930s. As for the 1938 Stocks, these were the best trains on the Underground *in* 1938. Today, they are quite corroded by the sea air, and the white paint is flaking off the roofs. The pleasing ripe-apple redness of the sides remains, although the cosy green of the original interiors has been banished, but the green is not so needed now. Its purpose was to suggest the countryside to London commuters, whereas on the Island Line there's real countryside beyond the windows. The trains have wooden window ledges, and a mixture of transverse and longitudinal seats, such as only remain on the Bakerloo of all the Tubes. They also still have guards, and those on the IOW are much more cheerful than the London ones used to be.

I alighted from the Tube train at Smallbrook Junction, where the Island Line connects with the IOW Steam Railway. Nothing else connects with the IOW Steam Railway at this point: Smallbrook Junction is accessible by rail only.

The Island Line station is a platform with a bus shelter-type arrangement, while the IOW Steam Railway station is a cricket pavilion-like thing with an open-fronted shelter alongside. The ticket clerk had propped a cardboard folder against the pigeonhole to keep out the flying rain. The train was also waiting. I boarded a third-class compartment in some bogie stock, the antiquity of which came home to me when I sat down on one of those extravagant, floral moquettes that characterise pre-grouping railways and that seem designed to be simply as pretty as possible. In *British Rail 1948–83: A Journey by Design*, Brian Haresnape wrote of early carriage design,

'For the decoration of carriage interiors it was often the wife of the General Manager or of one of his senior staff who chose furnishings and fabrics.'

I was in a silent wooden world, like a ship's cabin. 'Alarm Signal,' I read, beneath the communication cord. 'To stop the train, pull down the chain. Penalty for improper use £5.' A plaque below the string luggage rack explained about my accommodation: 'London Brighton & South Coast Railway composite number 7, built at Lancing in 1924 ... Transferred to the Island by the Southern Railway in 1924 ... This coach is the youngest on the railway.'

The case of the ex-Tube trains on the Island Line is not isolated. The island's railways have always had cast-offs from the mainland, and the IOW Steam Railway's rolling stock is exclusively of island pedigree, so it is *old*. Even before the days of railway preservation, the island was in effect a railway museum, and the entire IOW Steam Railway is actually classified as a museum. It began in 1966 when a group of teenagers, dismayed at the line closures, and the disappearance of steam from the Ryde–Shanklin line, founded the Wight Locomotive Society. They had half a dozen carriages and a locomotive: *Calbourne*, an LSWR O2-Class tank engine which is to the IOW Steam Railway what *Stepney* is to the Bluebell: a talisman or mascot. The preservationists began running *Calbourne* on the mouldering half mile of track between Havenstreet and Wootton (slightly north of the centre of the island). In 1991 the line was extended to Smallbrook Junction, where they could connect with the Ryde–Shanklin line. A further extension, into Ryde itself, has been projected.

We rolled away, dead on time – hauled by an Ivatt

Standard Class 2 tank built in 1952 (all the IOW Steam Railway's engines are tanks: nothing bigger has ever been required on the island). We moved through the misty, rainswept meadows. The sea was in the air, rather on the horizon. It was also in my mind, since a notice in the compartment advertised Southern Railway boat trains. But Havenstreet, the next station, and the HQ of the railway, is a rural rather than a maritime phenomenon. When the station was opened in 1876 by the Hythe & Newport Railway, it was connected to civilisation by a gravel track, so it was a haven *from* streets.

Havenstreet is not so much a station as a well-ordered complex in Southern Railway green. I had lunch in the café, which was packed with retirees. (This was the week before the Easter holidays.) At the servery, the cakes were all labelled: 'Isle of Wight Steam Railway Apple and Cinnamon Pie' etc., and the ingredients given. In the crowded dining area, a volunteer in an Isle of Wight Steam Railway sweatshirt directed me to a free seat. As I ate, the absent sea was the theme of my neighbours' conversation. One chap had grown up in Derby. '*Our* nearest sea,' he said, 'was Skegness – ninety-four miles away. And the joke was that when you got there, you still had to take a bus to sea, because the tide would be so far out.'

After lunch, I wandered into the second-hand bookshop. There were special book-carrier bags labelled Isle of Wight Steam Railway. The building housed a children's play area. I visited the gift shop, then walked to a big green hangar, opened in 2014 thanks to National Heritage Lottery funding and called Train Story. Here, the railway's impeccably kept vehicles are stored, so it's half loco shed, half museum. The railway owns all its own rolling stock;

nothing is leased from sub-groups as is the way on most preserved lines. A railway museum curator had commended the Isle of Wight Steam Railway to me by saying, 'It's a good little railway; they mow the banks,' and the area around the hangar was landscaped; a pond was being excavated. There was a slight air of being on a golf course. A 1950s beach hut, created from an old railway carriage, was exhibited nearby, with deckchairs set out in front. Inside, a Bakelite radio played *Hancock's Half Hour* on a loop. A notice read, 'Please note, the dials are glued and do not operate.' Not all of the old rolling stock taken to the island was used on its railways: 120 carriages were used as beach huts, hence this display. As I walked back to the main station, I read notices describing the maintenance by the railway of the local ancient woodland.

I congratulated a passing volunteer on all this excellent organisation.

'Thank you,' he said. 'The thing is, you see, we're very, very solvent.'

I asked why.

'Because of the tourism.'

I boarded a train for Wootton, the westerly terminus of the line. This is not the original Wootton Station; it was built in 1986. It's in wooden Edwardian style and liveried in the colours of the Isle of Wight Central Railway: two kinds of buff. As with the other stations, I could see no nearby road, and as we pulled away to return to Havenstreet, the volunteer guard, standing in the rain on the platform, cut a lonely figure, like a man trapped in the artificial past. I continued beyond Havenstreet to Smallbrook Junction, where I changed trains onto the Island Line. As I stepped on board the 1938 Tube car, the guard came up to me: 'This is yours, I think.' He held up the blue cap I had left on his train that morning. I liked to think he had recognised me. Or it might have been he was just looking out for a man with his unprotected hair plastered flat by rain. Either way, the transaction could not have occurred on a London Tube train.

The Swanage Railway

In Swanage, summer was being boldly anticipated. Buckets and spades (and, this being a resort of the traditional kind, I mean literal buckets and spades) were on offer outside many of the shops, along with beach towels, sun hats and so on. It was a day of weak sun, but strong wind, and the sea was leaping high at the harbour wall.

In 1885, someone connected with the Swanage Railway had told me, the railway came to Swanage – courtesy of the LSWR – 'and the town took off'. In 1972, the line closed, and the town threatened to crash-land. This was a late, post-Beeching closure: the axeman had marked 127 holiday resort stations for closure but somehow left Swanage off his list. So the town knew that a commutation of the death sentence might be possible. A preservation campaign was quickly in place, and it had something in common with the earliest preservation campaigns, in that the aim was a resumption of a full, year-round public service, connecting to Wareham on the main line, and so to Waterloo, from where seven trains a day had once come to Swanage, all year-round.

But the Swanage Railway Society had bad luck. BR would not relinquish the western end of the branch, because their tanker trains served an oil depot thereabouts – at Furzebrook – and without access to the main line, the local authorities would not at first support a restoration of the line. Without local authority support, the line was vulnerable, and BR lifted the track from Swanage to Furzebrook – that is, along almost the entire length of today's preserved railway. In *An Illustrated History of Preserved Railways*, Geoffrey Body writes that the Swanage Railway faced 'the full range of preservation obstacles', including a road development scheme, 'a problem no self-respecting preservation project should be without!' After negotiations too tortuous to go into, and the abandonment of the road scheme, the Swanage Railway Company (operational offshoot of the Swanage Railway Society) began running trains over a short distance of relaid track at Swanage in 1979. But it would not be until 1995 that the line was going 'from somewhere to somewhere': from Swanage on the coast to Corfe Castle, five miles inland. Access to the western end of the branch remained blocked. In having to relay miles of track, the Swanage Railway is what is known in preservation circles as a 'trackbed scheme'. Others include the Welsh Highland, which we have already visited, and the Gloucestershire Warwickshire, which we will be visiting shortly, and they all have incontestable bragging rights, the line guide to the Swanage speaking of 'the sheer dedication of thousands of volunteers and supporters who have strived for some 40 years to rebuild the track'.

For all its early travails, the Swanage Railway is today perhaps the best integrated into the national network of all the preserved lines. In 1997, the connection to Wareham was eventually brokered, and for the past couple of years, South Western Railway trains have run trains into Corfe Castle on summer Saturdays. These summer services have been inaugurated in each of the past two years with a direct service from Waterloo, and if, on some late-May date in the coming years, you see a 1950s-looking family on a platform at Waterloo – with actual suitcases, Dad with sandals and socks, Mum in a headscarf – they will either be ghosts, or people getting into the spirit of the Swanage Railway revival. In 2017, the Swanage Railway ran its own summer diesel service to and from Wareham, 'and lost money in the process', as one volunteer wryly informed me, even though there was a good take-up for the service and it is intended to reinstate it when a certain DMU has been overhauled.

The Swanage Railway is the centrepiece of Swanage's commercial life. The town had recently had a windfall when *Flying Scotsman* paid a visit to the railway, arriving under its own steam. 'All the roads round about were completely snarled up,' my Swanage Railway contact proudly informed me. *Flying Scotsman* does this wherever it goes, inflicting its revenge on the automobile.

The dependence of Swanage on its preserved railway seemed underlined when a man walked up to me as I was gazing at the thrashing waves and asked if I knew of a public lavatory. I pointed towards the railway station, which is visible from much of the town and audible from all of it: 'Sure to be one there,' I said, and when I myself ambled over to the station ten minutes later, I was pleased to see a typically capacious and gentlemanly preserved-railway Gents, this one done in BR Southern Region green.

The station evokes British seaside holidays of the 1950s and 1960s. In the booking office was a salacious poster from the second of those decades, showing a woman with a perm lying down on a beach and displaying her cleavage: 'Arrive Earlier by Train' ran the slogan. In the station shop, I bought a postcard that was either a reproduction of an old card or a pastiche. It showed a woman waving from a beach: 'Summer comes soonest in the South'. As I stepped back onto the platform, the wind blew the card away from my sheaf of papers. Just then a train came in, hauled by an air-smoothed 'light Pacific' engine (didn't catch the name) of the kind that typically brought the Waterloo trains to Swanage. The light Pacifics were designed by Oliver Bulleid, Chief Mechanical Engineer of the Southern Railway, whose creations tended to look strange. The light Pacifics suggested Spam cans, and that was their nickname. This one at Swanage hauled some green Southern Railway coaches, also by Bulleid, and their bulbous appearance was noted when we visited the Bluebell.

I spent most of the rest of the afternoon lounging in a first-class Bulleid carriage with a silver-and-blue Jazz Age seat moquette. I did alight at Corfe Castle Station, the principal one on the branch after Swanage itself. There is a small railway museum in an old goods shed, and the 'Corfe Castle Model Village', which is well made, but not really required, because Corfe Castle itself is a sort of full-sized model village, its trimness offset by the ruined actual castle on its hill. There were plenty of people about, walking between the village and the station, but they seemed to be going through the motions, slightly disengaged, like extras in a film waiting for the sun to come properly out, and the director to shout 'Action' – because you do need full sun in Dorset for that Famous Five vibe.

Aside from its main-line connection, there's another

way in which the SR is a 'real' railway. During the firework nights of the Swanage Carnival in late July and early August, a special late service operates along the line, with a 10.20 p.m. departure from Swanage. It would be nice to be on that train, tired from the heat of the day, the lamps lit in the first-class Bulleid compartment, the last of the fireworks floating down into the sea as the train rolls away towards one's car in the Park & Ride at Norden or, better still, one's bed in some hired cottage at Corfe Castle.

6

Specialists

Diesel

Diesel Romance

Diesels were not interesting. In *Pleasures of Railways*, Brian Hollingsworth wrote that, by virtue of its internal driving systems, 'A diesel loco looks very like an ordinary bogie carriage or van.'

The 1955 Modernisation Report was not quite the death knell for steam. Because Britain had more coal than oil, the Standards would continue to be built, but the push to dieselisation had become a rush, with inadequately tested designs prematurely commissioned. To the average railwayac, they did not look like the bringers of a bright new dawn; they were just ugly – and frequently unreliable – interlopers. In *Forget the Anorak: What Trainspotting Was Really Like*, Michael G. Harvey wrote, 'The introduction of diesel and electric locomotives and diesel and electric units severely depressed the majority of us local trainspotters.' In 1958, the Reverend Awdry put the boot in, presenting, in *Duck and the Diesel Engine*, 'Diesel', the only engine character in the series without a proper name. 'Sir Topham thinks I need to learn,' says Diesel: 'He is mistaken. We diesels don't need to learn. We know everything. We come into a yard and improve it. We are revolutionary.' ('I kept being pestered,' Awdry said, 'and I eventually agreed to introduce a diesel.')

The Diesel Train Driver, a British Transport Commission film of 1959, seemed to acknowledge the dreariness of the new mode. It shows a depressed-looking chap at the controls of a DMU. 'This is a diesel train,' begins the flat-toned commentary. 'There are several different models now in use. However, they are all generally speaking very much like each other. Diesel controls are very easy to use as they nearly all operate electrically ... Almost anyone can drive a diesel train after a short period of instruction.' There is then a cut to a gormless-looking teenager aboard the unit.

But it seems that the only thing required for a British railway phenomenon to become revered is the passage of time. In *BR Diesels in the Landscape* (1975), Derek Cross writes:

> The steam locomotive was undisputed master of the railway scene for a century and a quarter, but a mere seven years after steam vanished from the West Coast main line, diesels there have disappeared as well! Could it be that those of us who grew up in the Indian Summer of steam working and photographed a diesel if it came along, have captured a passing phase of railway history more important than we knew? A photograph of a Class 40 on an express climbing Shap may probably have as much historical interest in years to come as a Webb Compound taken there seventy years since, and it is a sobering thought that a 'Warship' in Wiltshire is no more repeatable than a Churchward 4–4–0 of half a century ago.

In books like *BR Diesels in the Landscape*, diesels are recalled with affection because they were the last engines to run on the doomed country branches: they went down with the ship. 'Today [the pioneers] would be astonished at the scale of railway preservation in Britain,' Peter Herring wrote in *Yesterday's Railways* (2002), 'and utterly perplexed that similar effort is now being put into saving the diesel locomotives they came to despise for usurping the role of steam after nationalisation.'

For the casual enthusiast, diesels are harder to love than steam engines by virtue of being more technically involved. All steam engines are powered by steam, but diesel engines might transmit their power to the wheels by one of three methods: mechanically (by a gearbox), electrically or hydraulically. Mechanical transmission is used on small engines (shunters, say) or on the early or more primitive units. Most of the big diesels are electrically driven, and the most powerful and glamorous diesel electrics were the Deltics or Class 55s, which roared up and down the East Coast Main Line (deafening some of their drivers), but, as with Gresley's streamlined A4s, which ploughed the same furrow, they were a rearguard action: Gresley was trying to meet the challenge of the motor car, while the guilty secret of the Deltics was that they owed their existence to governmental unwillingness to electrify the ECML.

The Western Region of BR, perpetuating the iconoclasm of the GWR, commissioned diesel-hydraulics, and these are the most affectionately remembered diesels: the Hymeks, Warships and Westerns, especially the latter, which were designed by Misha Black of the Design Research Unit, and which looked endearingly like Dutch barns on the move. At about the same time (mid-1960s) Black also designed the trains for the new Victoria Line of the London Underground, and these – which seemed to be wearing wraparound sunglasses – were considered among the most stylish Tube trains. The early Westerns pre-dated the tyranny of Rail Blue and had interesting colour schemes: one, D1000, was outshopped from the Swindon works in 'Desert Sand'; others were green or maroon.

In *Platform Souls*, Nicholas Whittaker writes:

> Exactly what a diesel-hydraulic was, I hadn't the foggiest, nor any inclination to find out. I didn't know what a torque convertor was (a Western had six, apparently), nor did I give a toss that the driving wheels were 3 ft 7 in as against the Warship's 3 ft 3 in ... I liked Westerns because they were maroon ...

Diesel-hydraulics can be seen on the East Lancashire Railway, the Severn Valley, the West Somerset and Midland Railway Butterley. The first of those railways is a diesel specialist, but we will be visiting it later under a different heading. The Severn Valley Railway offers mixed traction but has a big diesel gala every year, and we will be going there in due course.

But first a word about diesel fans. I spoke to one I met on the Epping–Ongar Railway: Chris Stevenson. He told me it was the withdrawal of the Western Region hydraulics in the late 1970s that really got diesel preservation underway. 'When the hydraulics started disappearing, people realised they had liked them. Then, in the mid-Eighties a lot of diesels were being replaced by units, so the old locos became available.' Part of the attraction of diesels was that they were cheaper than steam locos, as they still are: 'A typical diesel today might cost fifty thousand; a steam loco will be double that.'

Chris is a member of the Class 20 Society. The Class 20s – amusingly upright vehicles reminiscent of electricity sub-stations or toasters, and with a cab at only one end – were first introduced in 1957. 'They were basic freight locos,' says Chris, 'transferring wagon-load freight between marshalling yards, and of course a lot of that traffic went away pretty quickly.'

I suggested that this was a humble role, and Chris agreed. 'Yes, they're pretty unglamorous.'

I wondered if this was part of the appeal.

'I suppose so; the idea of these engines pottering between funny little freight depots ...' Chris, semi-retired from work in IT, conceded that diesel fans might be more technical than steam fans, the locos being more complicated, which in turn causes a maintenance problem that can limit the life of a preserved diesel. 'It's hard to get spare parts, whereas you can usually *make* a spare part for a steam engine.'

Diesel fans tend to be younger than the steam fans – or older, since most children visiting preserved railways prefer steam. (In a booking office of the Gloucestershire Warwickshire Railway, a father indicated his ten-year-old son: 'He's very intolerant of diesels.') If the typical adult steam fan is a seventy-year-old man in a V-necked jumper and tie, a diesel fan might be wearing a denim jacket, and if he orders a beer from the buffet car he might drink it straight from the can. 'We've had a bit of a problem on occasion with the diesel boys,' one senior volunteer on a preserved line that has mixed traffic (not the Severn Valley) told me: 'mooning out of the carriage windows, and so on.'

I stress that I personally have never seen any such thing (thank God), and that Chris Stevenson himself is one of the most gentlemanly characters I've met in railway preservation.

The Ecclesbourne Valley Railway

The Ecclesbourne Valley Railway began operations in 2002. It is part of a phase of preservations described in the railway press as 'new generation'. These lines were diesel specialists, aspiring not so much to steam dreams as the operation of a public service, which their diesels would be more likely to enable than a steam fleet. They sought close integration with the national network.

I saw this philosophy in action when I alighted from the DMU Sprinter, operated by East Midlands Trains, that had taken me from Derby to Duffield via the Derwent Valley line (which is not to be confused with the Derwent Valley Light Railway). On the other side of a footbridge, I saw another DMU, seventy years older than the Sprinter. A peak-capped platform guard seemed to be giving it the 'Right away', so I dashed over the footbridge. As I leaped aboard the old DMU, the platform guard, closing the door behind me, said, 'You needn't have run, you know.' The Ecclesbourne departures are timed to coincide with Network Rail arrivals, and if the East Midlands train is slightly late, or if a would-be punter is slow across the footbridge that connects the present with the past, the Ecclesbourne people will delay their train.

In this case, the train was a three-car Derby

Lightweight, which now got underway with a phlegmy cough and a belch of smoke. Derby Lightweight sounds like a Victorian boxer; in fact, the Derby Lightweights were the first class of the first generation of Diesel Multiple Units, whose introduction had been recommended in the BR Modernisation Plan of 1955. They were built at the Derby Works and tested on what would become the Ecclesbourne Valley Railway. It was said that people living near the line were spooked to see what appeared to be a carriage moving without the aid of a locomotive. The Derby Lightweights, like all early DMUs, were green, to match the colour of BR passenger locomotives of the time, although BR Engine Green wasn't quite the same as DMU green: it was lighter.

Derby Lightweights also had 'straw'-coloured speed whiskers painted on the front, which were meant to give an impression of momentum but are in fact reminiscent of an old man's handlebar moustache. The trains are elegant with large windows, and they *were* light, being made almost entirely of aluminium. The impression is of a gazebo on the move, or a leaf being blown along the line: no wonder people were so charmed by them when they first came in. They seemed to belong in the countryside, and could even masquerade as that yokel-ish phenomenon, the pick-up goods train,* if they were fitted with a GUV (General Utility Van). This one also had a restful green moquette which harmonised with the passing fields.

It was a warm and bright spring day, but the oil-fired heating was on, making for a nostalgic smell, until a senior

* A train that ambled along a country branch, collecting and depositing goods.

volunteer ordered one of his subordinates to turn it off. We would be travelling the 9 miles to Wirksworth, where the railway has its HQ and the line ends, unless you count the branch off the branch, which runs half a mile from Wirksworth up a 1-in-27 incline to Ravenstor and the National Stone Centre, a museum set amid what used to be five quarries. The primary branch exists because the Midland thought that a line through the Valley would enable them to avoid a section of track – the Ambergate Line – they were contractually obliged to share with London & North Western Railway. But soon after the Midland had begun building the line, the LNWR pulled out of the Ambergate, so it was no longer necessary for the Midland to run right through the Ecclesbourne Valley. The line stopped at Wirksworth, so becoming a branch. It closed to passengers in 1947 but some freight – limestone from the quarries near Wirksworth – was carried until 1989, when the quarries closed. But there was the possibility that they might re-open, so Wirksworth's station yard was given legal protection as a Strategic Freight Site. This special status prevented the track of the Ecclesbourne Valley Line from being lifted in the long hiatus before its preservation. It merely became extremely overgrown.

The preservation was under the auspices of a company called Wyvern Rail, which was created in 1992, when the privatisation of BR was in prospect. Wyvern Rail's first idea was to become a fully-fledged train operating company, running the line on a commercial basis. In the long wait for access to the line, this ambition was relinquished in favour of running the line as a standard preserved railway, with Wyvern Rail PLC operating alongside a supporters' Association. The clearing of the track began in 2000, and a lease-purchase with Network Rail was agreed in 2003.

The train guard came to collect my ticket, and I apologised that I hadn't been able to buy one. He sold me a ticket, pointing out that I could have bought one from whichever station I'd started from. That was St Pancras, so I looked at him doubtfully. It seemed a long way from St Pancras to the verdancy we were rolling through. He said, 'Well, we had a chap the other day who came from Ely to buy a car at Wirksworth, and he bought a ticket for Wirksworth at Ely Station.'

I said, 'Could you also buy a ticket to Wirksworth online?'

'I don't think you can do that, no.' (He's right; you can't.)

There are five stations on the line, all designed by the Midland Railway architect John Crossley and consisting of two pavilions with a central veranda. The Ecclesbourne looks like a slightly tamer version of the Settle–Carlisle line, whose stations were also designed by Crossley. Shottle, the station after Duffield, is owned by a company called Peak Oil. A sign on the platform reads, 'The station building is private. Passengers are kindly requested not to trespass on Peak Oil land or distract their staff.' But Peak Oil are obviously good sports, because this is an iron sign in the style of the old Midland Railway, and Peak Oil have painted the station in Midland Railway colours. Beyond Shottle, the line seemed to sink deeper into its valley; the banks became steeper and more overshadowed by trees. Where a vista did open up, rolling hills were visible some distance off. The landscape of the valley is restrained, restful rather than spectacular.

At the Wirksworth terminus – a pleasantly scruffy complex inhabited by preoccupied-looking men in high-visibility jackets – the senior volunteer called out that he was about to begin giving a tour of the station yard. It was rather a whirlwind tour, and my note-taking struggled to keep up ...

'That caravan belongs to our boss, because he lives in Southport ... That's an LMS sleeping car ... That's a little diesel-hydraulic with a Rolls-Royce engine; it was delivered to us by mistake ... That's a teak coach from the North York Moors Railway. It was vandalised last year, and we're repairing it ... That's a Class 08 shunter. Over there's the incline to Ravenstor – 1-in-27; we run trains up there on some weekends – sometimes steam trains, and it's our real money-spinner.' There was a boy of about thirteen on the tour, accompanied by his grandfather. Children of that age who visit the preserved lines either know almost nothing about trains or a very great deal. This kid was in the second category, and was rapidly becoming teacher's pet with his know-all questions: 'Is your Class 08 main line certified?' 'Isn't that an LMS sleeping car?'

After the tour, I had a cup of tea in a café that had been installed into an ex-Gatwick Express coach. When I turned up, a small group of volunteers were talking about ghosts, which seemed appropriate. 'He's seen it, she's *heard* it. Nobody has any explanation ... ' A man from this group told me that he was overseeing the installation somewhere nearby of a model railway layout. 'A chap gave us a nineteen-foot N-gauge layout. Two of our fellows knocked up the base board in about two hours.'

Not being familiar with Derbyshire, I asked the

volunteer how near we were to the Peak District.* 'We're just south of it,' he said, and I realised that the hills in the distance had been getting progressively higher since Duffield, as though the landscape were working itself up to something. 'I have a map somewhere,' said the volunteer, and he fished out a folder with the word 'INFORMATION' written on the cover in letters descending diagonally.

As I sipped my cup of excellent strong tea, I asked him whether he'd ever been on the Gatwick Express.

'Never been on it myself,' he said. 'I think this [the carriage] cost us about a pound.'

An hour later, I was back on the modern side of Duffield station, waiting for the train back to Derby. The clever kid was there with his grandad. A non-stopping train was approaching at a lick. 'Careful, grandad,' said the kid. 'A fast train can suck you onto the tracks.'

I don't think a Derby Lightweight would ever be so ill-mannered.

The Mid-Norfolk Railway

Like the Ecclesbourne, the Mid-Norfolk Railway was classified as a 'new generation' preserved line. Its original plan – so far unrealised – was to run a regular commuter service to Norwich via its main-line junction at Wymondham. It has several diesel locos and half a dozen DMUs. Steam engines sometimes appear, but they are always visitors from other lines. The Railway is based at Dereham, Norfolk, in a station that gives the impression of being besieged by cars, since its forecourt has become a huge car park, and there is another car park (belonging

*Home of the scenic preserved line, Peak Rail.

to a Morrison's supermarket) directly over the road. The plight of the station is made more poignant by its architecture, which is dreamily Jacobean, in the characteristic Great Eastern Railway style.

Dereham Station has been handsomely restored by the Mid-Norfolk Railway Preservation Trust, which took it over in 1997. It had been a victim of Beeching's particularly brutal assault on the railways of Norfolk: passenger services from the station stopped in 1969, with the ending of the Dereham–Norwich trains, and the station was gutted, becoming at one point a car showroom. The MNRPT acquired the route south to Wymondham (about halfway to Norwich) and re-opened that line between 1995 and 1999. The railway plans to extend 3 miles north from Dereham to the genteelly named village of County School, which would bring its length up to 15 miles. North beyond County School lies Fakenham, from where the line used to run to Walsingham then Wells, and these latter two places are joined today by the Wells & Walsingham Light Railway, which we have already visited.

The weather was right for the cinema verité of diesel traction: a grey Sunday afternoon, with cold rain threatening to become sleet. I was one of only half a dozen on the platform at Dereham, whereas on the Saturday, the man who sold me a ticket told me, the place had been 'heaving'.

Motive power was about to be provided by a throbbing early diesel, its front pleasingly rounded and tram-like, suggesting its late-1950s manufacture. This was a Class 26, a mid-sized or mixed-traffic engine, as built by the Birmingham Railway Carriage and Wagon Company, and appearing in Norfolk on this day as a visitor from the

Gloucestershire Warwickshire Railway. The Class 26 was in Rail Blue, but some of the 26s in preservation are in BR Engine Green, which they are entitled to be, given that they are sufficiently early.

An intent-looking man in grubby orange high-vis was climbing into the cab. I asked him about the appeal of diesel traction.

'It's as much part of railway history as steam,' he said, somewhat indignant, 'and *more* for anyone my age' ... which looked to be about *my* age (mid-fifties), and when I boarded the train for the run to Wymondham, I saw the truth of what he'd said. The carriages were BR Mark 2 – the carriages of my boyhood. Externally, they were in the utilitarian livery of 1970s BR, Rail Blue and grey. Inside, too, they were basically blue and grey. Essentially, colour was done away with on the Mark 2s, and so – gradually over the course of their production – were compartments. This was an open carriage, and at 3 p.m. on this gloomy Sunday I shared it with no other human, but a buzzing bluebottle. Viewed down the telescope of forty years, the carriage looked half-modern: yes, moulded plastic seats (a bit 1970s Conran Shop) but also wood panelling at the carriage sides and ends. The windows were partially open-able (at least the top panes), indicating no air-con. So these were *early* Mark 2s, dating from the late 1960s or early 1970s. The later Mark 2s relinquished the wood and open-able windows, becoming, according to Nicholas Whittaker in *Platform Souls*, 'as sterile as a DHSS waiting room'.

On the table before me was a Tunnock's Teacake, bought in the station café to go with my takeaway tea. Most of the preserved railways sell confectionary made by Tunnock's, probably identifying with its old-fashioned homeliness. Tunnock's are a 'family baker' at Uddingston in Scotland, and some of their packaging is branded with the Tunnock's Boy: an apple-cheeked kid with a neat parting. The firm is most famous for the surprisingly chewy Tunnock's Caramel Wafer, prettily wrapped in gold and red foil, which you will often see left behind (but neatly folded to postage stamp dimensions) on carriage tables of the preserved railways. (Strictly speaking, Tunnock's confectionary would look more at home on BR Mark 1s than 2s. I ought to have been tucking into some 1970s cuisine, such as a Casey Jones Traveller's Fare burger.)

We set off into some scruffy countryside. The engines of the 26s were built by the Swiss firm of Sulzer Brothers. There is supposed to be something called a 'Sulzer beat', but I detected only the usual soporific and not unappealing diesel drone. Another drone was going on above my head: the trapped bluebottle. I thought of it as a dazed leftover of the 1976 heatwave.

The train guard came up, a friendly and amusing chap in his seventies. 'I've done all sorts on this railway,' he said. 'I was in personnel training; then I was on the crossings, then I was a guard, then on the computer system, arranging shifts. Then I was back to guarding.' Seeing that I was making notes, he offered a summary: 'We're all nuts!' He went away but came back a few minutes later, having had a second thought: 'The word is "enthusiasts".' We happened to be rolling past some general trackside lumber that included a pile of wooden railway sleepers. The majority of railway sleepers on the preserved railways are wooden, as opposed to the more modern version in concrete. 'They used to be five pounds apiece,' said the guard, 'but then Alan Titchmarsh and Charlie Dimmock recommended use in gardens, so they went up to forty pounds.'

After half an hour, we came to the countrified halt called Wymondham Abbey, which is the southerly terminus of the MNR. It is not to be confused with Wymondham Station a mile away, which lies on the Breckland Line between Cambridge and Norwich, although the two Wymondhams are rail-connected, which is how the Class 26 got to the Mid-Norfolk from the West Country and how, one day, the Mid-Norfolk might connect to Norwich. It's a shame that there's no room for the MNR at Wymondham Station, since it's more like a preserved railway station than most preserved railway stations, being smothered in flowers all summer, and featuring a bistro (where you sit on carriage seats – first-class ones) and a small museum. At Wymondham Abbey there was a cubicle-like ticket office-cum-waiting-room-cum-second-hand-bookshop.

As the Class 26 ran around its train, I enquired about the timings of the services back to Dereham. I could either go back there in ten minutes or in an *hour* and ten minutes. I walked into the historic market town of Wymondham, which took two minutes. In the next two minutes, I ascertained that everything in the town (including the Abbey itself) was closed, so I walked back to the halt, and joined half a dozen others admiring the 26 – now at the right end of the train – before riding back to Dereham. It had been an enjoyable but slightly disorientating afternoon. Later, I diagnosed the reason. I'd had a daydreaming, country-branch sort of afternoon (the empty train, the trapped fly, the somnolent market town), but without the motive power (steam) or the weather (sunny) usually associated with such a thing. I had not had a generic experience because, while old diesels are evocative, they're not quite part of our national mythology. They evoke particular memories in particular people, not one standard folk memory.

Diesel Gala on the Severn Valley Railway

Kidderminster Station on the Severn Valley Railway appears to be the main station of the town, being a substantial Victorian-looking building. It stands next to something resembling a small modern bus station. In fact, this is the town's Network Rail station, and it is probably not much older than the station of the preserved line, which was constructed 'to a traditional Great Western turn-of-the-century design' in 1986. It stands on the site of the first Kidderminster station, which was built in 1852 in mock-Tudor black and white, and which died of dry rot in 1968.

The Severn Valley line was a 40-mile-long branch of the GWR closed by Beeching in 1963. In 1965, the Severn Valley Railway Society was formed with the aim of preserving 16 miles of the line, between Kidderminster and Bridgnorth. With money raised by the flotation of a public company, the SVR gradually acquired the line, reopening a first stretch in 1970. The line was opened along its full extent in 1984.

Behind the Kidderminster Station façade is a glazed-over forecourt, and it was bustling like a market square. There was a green wooden WHSmith kiosk, acquired from Manchester Victoria station in 1986 (so much more relaxing than a modern Smiths), with historic headlines on newspaper placards: 'Malcolm Campbell sets 301 mph Land Speed Record.' There was a 'Station Emporium', and many stalls selling railway artefacts, the centrepiece being a trestle table covered with shoe boxes full of thousands – perhaps tens of thousands – of postcards of diesels. Several propped-up blackboards announced the serving of tea in the station pub. It involved beer.

The Severn Valley is mainly a steam railway, but it has the resources to lay on a big event, and the main-line junction at Kidderminster to bring in guest diesels, hence the Spring Diesel Gala. An article in that month's *Railway Magazine* had previewed the line-up, and the visiting locomotives would include a Class 31 from the Chinnor & Princes Risborough Railway, a Class 40 from the Class 40 Preservation Society, a Class 44 from the Peak Locomotive Company and a Class 55 (*Royal Highland Fusilier*) from the Deltic Preservation Society. I cannot pretend this meant a great deal to me, apart from the Deltic; I think I am more interested in diesel enthusiasts than diesel locomotives.

I boarded the waiting train. The engine was *Greyhound*, a Warship-class diesel-hydraulic in maroon livery. 'It was meant to be D1062, but it's all got discombobulated,' a bloke was saying as I boarded. (D1062 is a Western, also a diesel-hydraulic.) The carriage was older than the locomotive, being, as a notice taped on the window announced, an 'LMS 1936 buffet car' – all sombre Rexine and chrome, in which context diesel blokes (and there were no women in the packed compartment) seemed slightly displaced. They were mostly in their fifties. Some, I suspected, were ex-punks. One bloke was all in black; another wore a T-shirt reading, 'The Stranglers: the Definitive Tour'.

Their natural home would have been a BR Mark 2 carriage of the 1970s, but they'd landed on their feet here, because this was the buffet car, and most of them were drinking pints. I had arrived late at the Severn Valley, by the way, and it was now mid-afternoon. It takes a couple of hours to go along the line to Bridgnorth and back, and when I suggested to my neighbours that I might get off beforehand, one said, 'You've got the Engine House café at Highley, that's fully licensed.'

Another said, 'You've also got the Ship Inn at Highley – that's a lovely spot, right on the river.'

But another suggested I hold on for the end of the line: 'You've got the Pilgrim's Arms at Bridgnorth – that's worth seeing.'

The man sitting opposite was keen on smoking as well as drinking, and he was preparing neat rolls-ups from a tin that had formerly contained Altoids mints. The man on the opposite side, whose T-shirt read 'Still Got the Blues', was accompanied by his son, who was looking at pictures of diesels on his iPhone.

We were rolling by now towards Bewdley, where it seemed as if the town itself was preserved, with its Georgian architecture and the flag of St George fluttering from the church tower.

I got off at Highley, halfway along the line, and when the train pulled away I was suddenly back on the Victorian GWR, which is the default setting of the SVR. I suggested to a platform guard that this flower-bedecked station was in the famous GWR livery of light and dark stone.

'Yes,' he said, 'plus the *actual* stone.' One of his colleagues, who can't have been more than about eighteen, wore a black silk waistcoat and had a corncob pipe in his mouth; I don't think the pipe was lit.

I boarded the next train back to Kidderminster: this was hauled by a Class 50 diesel, namely *Hood*, an SVR resident. At the next stop, Arley, a group of lads boarded: trainspotters – and diesel specialists. They sat down on the opposite side of the aisle to me, one of them apologising profusely when he bumped against my legs. They were nice lads. Asking if it was OK by me, they opened all the windows on their side, then they sat down at a table. They were in their late teens, all good-looking in a Byronic way, but their longish, bouffant hair would make them slightly uncool amongst their peers, I suspected. They all carried sophisticated electronic cameras, and were quite well-dressed as well – no anoraks (not that anoraks exist any more), although one of them wore a sweatshirt emblazoned with the name of the defunct freight operator, EWS, which used a lot of diesels. They were planning the rest of their day: 'Peak up to Arley,' one of them said, 'then bail off for a Crompton, then bail on the 37 or 42. Then we rest, maybe food if needed.' (Peaks are Class 44, 45 or 46s, and called Peaks because the earliest Class 44 locomotives were named after mountains; Cromptons are Class 33s, named after their Crompton Parkinson electrical equipment.)

One of the lads was showing images on his camera to a colleague. 'That's the 31. Have you heard the 33?' He played the sound of the Class 33; they all listened in an apparent state of bliss. This was perpetuated when another diesel went past: they all inhaled the fumes deeply, like the Bisto Kids. This was why they'd opened the windows. 'It's hellfire!' one of them said. Apart from 'bail' (meaning to board or alight from a train) 'Hellfire' was their favourite word, along with 'Dreadful', meaning the opposite of dreadful. (When I returned home, I read on a railway forum, 'Hellfire, Dreadful and My Lordz are all just basically 70s/80s-style yells from a basher who has got lucky with a rateable loco or just had a particularly nice ride behind something special.')

Back at Kidderminster I spent an idyllic half-hour having tea and cake in what was a cross between a café, a museum, a library and bookshop. I sat next to a boy of about five and his mother.

'What's best?' she asked the boy: 'diesels or chuffa-chuffa steam trains?'

'Chuffa-chuffa steam trains,' he said.

'Yes,' said his mother, 'but you don't have to get up at dawn to start the fires.' She was obviously a rather complicated and interesting person. I hoped that she and the boy had not had a disappointing day, given the absence of chuffa-chuffa steam trains, but they could always come back for the next day of the Gala, when absolute dieselisation would give way to 'mixed traction', or diesel and steam.

As I stepped out of Kidderminster Station, I saw a small, battered car – a Nissan Micra. A transfer on the top of the windscreen took the form of a ball of flame; underneath was the word 'Hellfire!'

Freight and Industrial Railways

*The Romance of Freight and Industrial Railways**

Towards the end of my day on the Great Central – as described in the Introduction – I was taking a photograph on the rainswept field of ash (a former goods siding) adjacent to Quorn & Woodhouse, where the 'swapmeet' had been held. All the stalls having been packed up and cleared away, I became conscious of what had been the backdrop all along: a line of dozen pale grey coal wagons, marshalled on the one remaining siding. Parked adjacent to them was a rusting Morris Minor of exactly the same pale grey, with a faded orange pennon-shaped sticker on the rear window reading 'I've been to Mablethorpe' above an image of a setting sun. The only other person left on the railway, it seemed, was the signalman in a nearby tall box, who stepped onto his balcony and looked enquiringly at me.

'Do these wagons ever move?' I shouted over to him, through the wind and the rain.

'A couple of times a year,' he shouted back, 'we do a demonstration coal train.' Over further shouted conversation I gathered that the coal wagons were not filled with coal for these events: they just rumbled empty through the stations for the benefit of spectators and photographers.

There is – and was – a romance attaching to freight trains. In my mind, the twilit environs of Quorn & Woodhouse were becoming like Mugby Junction, as described

---

*In our historical context, we should really be speaking of 'goods', which is the English term, rather than 'freight', which is American. But 'freight' gradually displaced 'goods' and has predominated since the formation of BR.

in Charles Dickens' short story of that name. Here is Mugby Junction shortly after 3 a. m. on a 'tempestuous morning':

> A place replete with shadowy shapes, this Mugby Junction in the black hours of the four-and-twenty. Mysterious goods trains, covered with palls and gliding on like vast weird funerals, conveying themselves guiltily away from the presence of a few lighted lamps, as if their freight had come to a secret and unlawful end. Half-miles of coal pursuing in a Detective manner, following when they lead, stopping when they stop, backing when they back.

The railways were built for freight, and freight traffic remained more important than passenger traffic until the mid-1960s. So it might be thought that the real railway work was carried on in the goods yards and goods stations, and most of it was carried on at night, or in mysterious zones off-limits to passengers. See Dannie Abse's poem, 'The Shunters':

> *Rain drags darkness down where shunters work*
> *The blank gloom below hoardings, dejected sheds,*
> *below yellow squares in backs of tenements*
> *whilst, resigned, charcoal trucks clash and jerk*

Another romantic aspect of freight trains is that they have the honour of being victims: first of cars, secondly of Beeching. BR was hampered in its attempts to compete with road hauliers by the 'common carrier' obligation imposed on them by the Railway and Canal Traffic Act of 1854 in an attempt to prevent the railways abusing their monopoly on freight traffic. An engine driver on the West Somerset Railway, who had also driven for BR, and whom I talked to as he sat in the cab of a DMU at Minehead, was sill embittered about this in 2018: 'If you'd wanted to move a giraffe from Cardiff to Aberdeen, the railways would have done it for you, because they had to do it. They'd have done it bloody well, as well, and at a reasonable price fixed by government, whereas the lorry drivers could charge what they liked, and carry what they liked ... And in the case of the giraffe they'd have told you where you could stuff it.'

(I think I know why the giraffe popped into his mind. In the late 1960s Hornby Tri-ang offered to railway modellers a 'Giraffe Car': the head of a giraffe protruded from the top of a wagon, and it worked in conjunction with a catch, so that the head of the giraffe would duck down when the train went through a tunnel. Hornby Tri-ang, a very imaginative company, also produced an 'Exploding Car': 'Hit this lethally laden freight car with a missile from either of the Tri-ang Railways Rocket Launchers and it will explode most realistically.')

The common carrier legislation would not be repealed until 1967, by which time Dr Beeching's massacre was under way. Those determined to say something good about the man point to his streamlining of BR's loss-making freight operations: the replacement of 'wagonload' with 'trainload', by which diverse trains of sundry goods were displaced by uniform trains of containers all carrying the same thing. This strategy was only partially successful in fending off road competition, but it bore the Beeching trademark of dullness. He brought an end to the nocturnal theatre of the marshalling yards: men assembling trains like cowboys herding cattle; walking or running alongside wagons that had been deliberately set rolling down an incline. It was enthralling to watch this dangerous work being carried on under floodlights at Dringhouses Marshalling Yard, York, when I was returning home after a night of under-age drinking; and when I was home in bed, it was still going on. In *Yesterday's Railways*, Peter Herring wrote of the marshalling yards: 'Off-key clanging, like the peal of cracked church bells, stood for the city at night.'

A category overlapping with freight is that of 'industrial' railways: colliery, quarry or factory lines, often privately owned and narrow-gauge.

The last working industrial steam locomotive was bringing coal to Castle Donington power station as late as 1988, over lines that are now an industrial estate. The four-coupled saddle tank engine concerned, *Castle Donington 1*, was the star of a BBC documentary in the *Train Now Departing* series of 1988. The driver, a Pickwickian man with a Dickensian name, Lionel Gadsby, was filmed on the footplate: 'As regards steam,' he says, in a Midlands burr, 'I think we are the last.'

The narrator then intervenes, 'The last to use steam to bring the coal, to make steam', a reminder to progressives that while steam might have been coming belatedly to an end on the industrial railways, it was still generating our electricity.

'As regards working conditions,' Mr Gadsby resumes, 'actually ... you're up to your neck in filth, coal dust, all elements, bad weather, rain, draught, but it's a lovely machine.'

The engine went into preservation, firstly at the Midland Railway Butterley, then to the Colne Valley Railway in Essex.

An important preserved industrial line is the Chasewater Railway in Staffordshire, with about thirty steam and diesel locomotives, which run over 2 miles on a former colliery branch line that closed in 1960. The genesis of the Chasewater is also important in relation to the wider history of railway preservation.

What eventually became the Chasewater Light Railway arose from a meeting held at the LNWR Hotel in Stamford on 16 November 1959, which formed the West Midlands Railway Preservation Society. Presiding was Noel Draycott, who envisaged the West Midlands group as one of a number to be created under the aegis of a national Railway Preservation Society, which Draycott had founded, also in 1959. 'This was to be a national society with a broadly federal structure,' writes Jonathan Brown in *The Railway Preservation Revolution*, 'having a number of district groups responsible for collecting historical material from their regions'.

Things did not always go smoothly for the West Midland Railway Preservation Society. I quote from the website of the Chasewater Railway:

> The West Midlands District had a couple of disappointments which at the time seemed major, but in retrospect now seem amusing. These included the theft by 'tatters' of a gentlemen's cast-iron convenience from the closed Stowe-by-Chartley station, for which BR had been paid £12. A venture which resulted in a financial loss of over £100 (a large sum

then) was the hiring of *Flying Scotsman* for a return Sheffield Victoria–Marylebone excursion on 15 June 1963. This was probably the first hire of the loco by the preservation group. One unexpected bonus in Sheffield was the donation of one of the LNER stainless steel *Master Cutler* headboards.

The society 'stuck to its task', acquiring semi-derelict locos and gaining a lease on its branch line in 1964. Gradually it became an operating railway rather than a museum, and it was in 1973 that the West Midlands Railway Preservation Society became the Chasewater Light Railway Company. There were still setbacks – 'some incredible vandalism occurred' – and the railway closed to the public for three years from October 1982 'when a miserable wet Saturday saw just two fare-paying passengers.'

Only two other societies prospered under the Noel Draycott scheme. One was the London Railway Preservation Society. This became the Quainton Railway Society, which runs the Buckinghamshire Railway Centre at the closed-down Quainton Road Station. The other was the Scottish Railway Preservation Society, which set up the Bo'ness and Kinneil Railway on the Firth of Forth from 1979. Ten years earlier, the SRPS had been approached by the new Highlands and Islands Development Board to preserve the line from Aviemore to Boat of Garten. It dropped out of the scheme in 1971; a Highland line was too remote for an organisation based at Falkirk. But six of its members broke away to form the Strathspey Railway Company, and we have already visited the line they created.

Draycott's national Railway Preservation Society had

gained a rival in 1962, in the form of the Railway Preservation Association, founded by Captain Bill Smith (that pioneer locomotive purchaser), but that didn't flourish either. In 1965, the National Council of Draycott's RPS reconstituted itself as the Association of Railway Heritage Societies, and the catalyst here was another Captain: Captain Peter Manisty. The ARPS would be a trade association for heritage railways, and as such was the forerunner of the Heritage Railway Association. Manisty developed good relationships with BR, and with successive governments, and it was he who brokered the sale of so many locomotives from Dai Woodham's scrapyard at Barry to preservationists.

While pursuing our joint theme of preserved railway national organisation and industrial railways, we should mention David Morgan. He is a dapper and energetic solicitor, usually seen in a Derby hat, and was a long-time president of the Heritage Railway Association and preserved line benefactor. He has advised many preservationists on legal matters. 'Level crossings are especially fraught,' he told the *Law Society Gazette* in 2014, 'because when they are mismanaged, the results can be fatal.'

Morgan's name came up when I was reading the website of the Heritage Shunters Trust. Here is a similar tale of underdog, 'Tough-of the-Track' persistence to that of the Chasewater, and David Morgan helped keep its industrial flame burning by finding it a home on the freight yard of an ex-Great Central branch from Meadowhall to Barnsley. 'David came to our rescue Big Time!!' reports the website. But the Trust was plagued by thieves and vandals on its branch line. The Heritage Shunters Trust is now more securely located, with a shed and running line for its dozen ex-BR shunters, at Peak Rail's Rowsley Station.

(Morgan, incidentally, was once approached by a volunteer on a preserved line who confessed that, many years before, he had absconded from an open prison, and he was being blackmailed about this by a policeman who had recognised him. Morgan arranged a sting: he advised the volunteer to meet the policeman in a pub, where his extortionate demands could be overheard by some un-bent coppers. The plan worked: the blackmailer was arrested. The volunteer was returned to prison but given full remission.)

Rocks By Rail

'The Midlands are a complete mystery to most people.'

So said a volunteer at Rocks by Rail, which has preserved part of a network of ironstone quarry lines in Rutland. The network was connected to the 2-mile Cottesmore branch of what had been the Midland Railway, north of Oakham (which survives as the only passenger station in Rutland). The volunteer made the observation about the Midlands when I said, 'I didn't know ironstone had been mined in Rutland.' The volunteer was himself originally from the West Country. He was a retired architect, who'd moved to Rutland and become involved in Rocks by Rail because he was able to help with the conservation of some engineering drawings. I told him that the Cottesmore branch was shown on the generally excellent *Railway Atlas Then and Now* as being 'closed to all traffic as of 1 Jan 2012', and he shrugged as if to say that this only proved his point. The volunteer instanced a typical enigma of the Midlands: 'Rutland is the smallest county in Britain when the tide is out, but when the tide is in the Isle of Wight is the smallest county.' (The Isle of Wight being bigger when the tide is out.)

There were three quarries on the Cottesmore branch, half a dozen others elsewhere in Rutland, and many more in the wider Midlands. This is, according to the Rocks by Rail website, 'very much an untold story for the young, new residents and visitors to the area', and that's probably true of ironstone mining in Britain generally, which ended in 1980 when the British Steel Corporation works at Corby switched over to using higher-grade foreign ironstone. But there is still, according to the volunteer, 'plenty of ore in Rutland and Lincolnshire.' ('The destruction of this once vast industry of Midland ironstone extraction', Colin Garratt writes In *British Steam Nostalgia*, 'by the claim that it is cheaper to import ore from overseas, is regarded by many who worked in the industry as dubious, if not actually sinister.')

Rocks by Rail was formerly the Rutland Railway Museum; it changed its name to something catchier on the advice of the local authority. The focus for some visitors will be the prefab engine shed and adjacent loco yard containing about twenty industrial locomotives. For the less technically minded, the focus will be the white breezeblock interior of the Sundew Café, in which delicious bacon butties are served with strong tea. The caff seems like a cross between a school room and barracks. There are maps and notices on the walls – the latter illustrating the harshness of quarrying life. Browsing with mug of tea in hand, I read:

Dangerous Quarries: trespassers will be prosecuted.

Treatment of Electric Shock: A person apparently dead may be revived by the method described below.

Accident or illness: Dial 999 for ambulance. Phone Dr Bradford Market Overton at Thistleton 229. Inform mine manager.

At the far end of the room, I came to a papier maché model of a quarry in which sat a model of an enormous excavator. This is – or was – *Sundew*, and notwithstanding that impressive collection of locos in the engine shed, *Sundew* is the presiding deity of Rocks by Rail. *Sundew* was manufactured by the superbly named Ransomes & Rapier of Ipswich in 1957, and named after that year's Grand National winner. Its job was the removal of the limestone overburden beneath which ironstone lurks. Ignoring the 300-foot reach of its jib, it was about the size of a squarish four-storey office block. Every bite of its bucket picked up 30 tons of rubble, and forty people could stand within it. It worked in one of the local quarries until 1974, when it relocated to a quarry near the British Steel-run ironstone quarry near Corby. *Sundew* relocated itself by walking – paddling or waddling, really – over 13 miles of snowy fields at one tenth of a mile an hour on its electrically powered feet. The journey took eight weeks, and there is film of the perambulation on YouTube.

Sundew was scrapped in 1987, but its cab was saved and restored with Heritage Lottery money. If you step out of the Sundew café and follow Rocks by Rail's short railway line in an easterly direction, you come to the cab, which is about the size of a suburban summer house. It overlooks a mock quarry where much smaller earth-moving machines sometimes perform in conjunction with quarry trains. On more normal operating days, trains simply shuttle up and down the half-mile line. Passengers, travelling in an old guard's van, are propelled in the direction of the old junction with the Midland by ... well, on the day of my visit, it was a four-coupled saddle-tank engine built in 1940 by the firm of Andrew Barclay and named *Sir Thomas Royden*, after the chairman of the power station where the engine spent the entirety of its pre-preserved life.

The fireman was called Ross; he was nineteen. He would not be allowed to drive until he was twenty-one, but he certainly *would* be driving. Ross worked as a plumber. He got his love of railways through his grandfather, who was 'into five-inch-gauge railways', and worked as a long-distance lorry driver, whereas Ross's father was a businessman: 'My dad's not technical; it skipped a generation.' In summer, Ross comes to the line at 6 a.m. to fire up a loco and leaves at 6 in the evening.

We went along the line, running under trees or past hedges. The retired architect, who was riding with me, said, 'We have one volunteer who does nothing but cut back the branches here.' Rocks by Rail plans an extension. It will only take the line a little further, into a copse of trees, but a platform will be created, so that the trains have a proper destination rather than the parked ironstone wagon that marks the current terminus. I asked the retired architect whether a certain melancholia was meant to be part of the appeal of the preserved industrial railways.

'Oh yes,' he said, 'absolutely.'

The Leighton Buzzard Narrow-Gauge Railway

The Leighton Buzzard Narrow-Gauge Railway is a preserved industrial railway celebrating the sand quarries that existed, and to some extent still exist, around Leighton Buzzard in Bedfordshire. It runs for 3 miles between Pages Park and Stonehenge Works, which are the only two stations. Terry Bendall, chairman of the railway, asked whether I knew that the glass of the Crystal Palace had been made from Leighton Buzzard sand, or that Leighton Buzzard sand is exported to Saudi Arabia. I did not. (The Saudis, apparently, have 'the wrong kind of sand' for some industrial processes.)

The railway was started as the Leighton Buzzard Light Railway in 1919, connecting various quarries with exchange sidings for the West Coast main line adjacent to Leighton Buzzard main-line station. During the First World War, the demand for sand had rocketed. It was transported by steam lorries, which ripped up the unmetalled roads, but Bedfordshire County Council was willing to pay for the road repair, given the emergency. After the war, the hauliers were told that from now on they would have to pay for any damage they caused, hence the idea of a sand-carrying railway. Some of the track and rolling stock for the Leighton Buzzard Light Railway came from the Western Front, where – from 1916 – temporary 2-foot-gauge lines had been laid down by the War Department Light Engineers to carry ammunition and men from behind the lines to the forward trenches.

In a yard outside its easterly terminus, Stonehenge Works (named after a type of brick supposedly as durable as Stonehenge), the railway has some rusting track lengths that actually were used on the Western Front. It also has a steam locomotive used on those lines. This was made by an American firm, Baldwin of Philadelphia, and is numbered 778. There are no bullet holes in 778, but the railway possesses a movement order for the engine dating from 1918, and mounted on the wall of the engine shed at Page's Park is a photograph of a row of Baldwins inside an engine shed somewhere in France near the front. With the use of a magnifying glass, it is possible to discern that one of the engines carries the number 778. Internal combustion engines were also used on the Western Front, but when, in 2010, I was researching a novel set on those railways, I was interested in the steam engines, which (a decadent consideration given the context) I regarded as more picturesque than the diesel or petrol ones. That was when I made my first visit to the LBLR – for a ride on the footplate of the Baldwin.

On that occasion, I was so busy trying to imagine myself rolling towards the forward positions under a hail of shells that I quite blanked out the actual setting of the railway, which for most of its length runs between the back gardens of suburban Leighton Buzzard.

The driver and fireman who had kindly allowed me into their cramped cab had explained some of the idiosyncrasies of the Baldwins' performance in wartime conditions. The engines, with their tall chimneys and high-mounted side tanks, had a high centre of gravity, and often toppled over. The track should take some of the blame. A mile of it might be laid down literally overnight, and it was sometimes laid directly onto mud, with no ballast.

The railway also owns a couple of the long wagons in which shells or soldiers were taken to the front, the men standing up in the wagons, as toy soldiers are obliged to do when played with in conjunction with a train set. Soldiers brought back in the other direction often couldn't stand, and planks were fitted across the tops of the wagons for the placement of stretchers. Sometimes the Leighton Buzzard people hold commemorative First World War events, but the main focus of the railway is sand.

By the early 1960s, much of the sand was once again going by road, and the railway was in decline. It was then that (in the words of Terry Bendall) 'Two chaps from St Albans came to the line.' They wanted to buy track and rolling stock for a pleasure railway at Watford, but instead they came to an arrangement with the owners of the LBLR to run pleasure trains for passengers along it at the weekends, while the sand-hauling business continued on weekdays. Enter our old friend Dr Beeching, who closed the connection between Dunstable and Leighton Buzzard, thereby removing the LBLR's exchange sidings at Leighton Buzzard, and leaving the line high and dry. But the LBLR never closed. In 1969, the enthusiasts – who had recruited a team of volunteers – leased the line from the Leighton Buzzard Light Railway company, and changed its name to the more informative Leighton Buzzard Narrow-Gauge Railway, which now became entirely devoted to pleasure and reminiscence, at first under the control of the two chaps from St Albans, whose motives were altruistic throughout.

My second visit to the railway – for my meeting with Terry Bendall – occurred on a crisp, cold day just before Christmas, and 'Santa Specials' were being run (see later: *Christmas Days*), so three things were contending somewhat unharmoniously in my mind: sand, the First World War and Christmas.

After a quick ride on one of the Santa Specials, Terry Bendall offered to show me the engine shed at Pages Park. Terry – the only volunteer not wearing either a Santa hat or a Christmas jumper – is a retired metalwork teacher, and looks the part, being a big man with big, capable-looking hands. He put me in mind of an intelligent blacksmith. He first joined the railway in hopes of 'doing a bit of machining', but he was deflected into the personnel side of things. He never has done any machining for the railway, and it was with a certain wistfulness that he showed me the powerful lathes and other kit in the machine room adjacent to the engine shed. 'Tuesday is our engineering day. We'll have ten or twelve people in here then. They might be doing anything from cleaning out boilers to making new axles for carriages.' (The railway makes its own carriages.) I pictured the roomful of retired blokes, but Terry said that the railway is not badly off for young recruits, and those who come forward tend to stick around. 'Our steam engineer started at the age of ten. He's now thirty-five.' (In the early days of the line's pleasure use, Terry said, it was a playground for some children. They'd get in a sand skip and ride it towards the buffers on the downhill stretch, then jump out before it hit. 'Wouldn't happen today,' said Terry. 'Health and Safety means you'd be sued if the kid got hurt.'

The engine shed at Pages Park doubles as a static exhibition called 'The Engine Shed Show'. The LBLR began to use internal combustion machines from 1921, having found that sand got into the motion of steam engines, but the steam-driven Baldwin, 778, was in the shed, undergoing restoration. After war service, it worked on a military line on the North-West Frontier, then at a sugar mill in Uttar Pradesh which, according to the written guide to the Leighton Buzzard Narrow Gauge Railway, 'became a place of pilgrimage for photographers and locomotive enthusiasts from Europe, including some of our own members'. 778 was brought to Leighton Buzzard thirty years ago, and a fund was established for its restoration.

Small locomotives are still being manufactured for industrial use, and the engine shed accommodates a diesel that was used in the building of the Channel Tunnel. But most industrial sites now use tracked vehicles – vehicles that (you might say) roll on their own rolling railway lines in the form of caterpillar tracks. They are almost always yellow.

Terry drove me to the other terminus of the line: Stonehenge Works. As we sped past progressively newer houses, he said the railway would be extended a little way beyond Stonehenge Works to escape the ever-expanding suburbs of Leighton Buzzard, and give passengers half a mile or so in the countryside. He acknowledged that the railway would never be 'pretty like the Ffestiniog', but the new stretch ought to give views of 'deer in the fields and hawks in the sky'. Adjacent to the station at Stonehenge Works was a big worked-out quarry, looking like a sandy hole about the size of a football stadium. Within the station yard was a hill of sand with a couple of early 1960s diesel excavators (or 'face shovels') alongside. They were cream, maroon and green, and so more cheerful-looking than their jaundiced, hydraulic modern equivalents. On 'special occasions' they give demonstrations, loading sand into sand skips which roll around a small loop of track.

As I looked down the line towards Pages Park, the sun was sinking over the new-build semis. Terry Bendall and I stood in a deserted children's play area, where a small, faded tank engine functioned as a climbing frame. The engine had a name: *Penlee*. It came to Leighton Buzzard in 1991, and the guide to the line says, 'Despite her very antiquated appearance, she would be quite powerful for her size if restored to use, but her very limited water capacity would make operating difficult.' As Terry Bendall and I walked back to his car, he told me a story about *Penlee*. The engine was built in 1901 by the firm of Freudenstein in Berlin, and first worked for many years on a line between Penlee Quarry and Newlyn Harbour in Cornwall. Initially, the engine was called *Berlin*, a name that became intolerable in the First World War, so *Berlin* became *Penlee*. For forty years, *Penlee* was always driven by a Mr Jenner, who always wore a bowler hat. When the engine was withdrawn from service in 1946, Mr Jenner was off work with a serious illness. He was so upset about the withdrawal of *Penlee* that management told him a white lie to the effect that it was only temporary and that they were awaiting new parts from Germany. Mr Jenner was consoled by this thought in the final months of his life.

The Watercress Line

That the Mid-Hants Railway is media-savvy is surely suggested by the fact that it has a marketing name in addition to its formal one: 'The Watercress Line'. The railway also employs a media officer, and I tend to think that a line having one of those can afford to give me a press ticket to go up and down the line, so in my first summer of writing this book, I applied for one: no reply. In my second summer I called the line, and the volunteer who answered the phone agreed to put my question to the media officer. He came back a minute later: 'The answer's no.'

The Watercress obviously reasons that it doesn't need any extra publicity from the likes of me, and when I arrived at the main station and westerly terminus, Alresford, the car park was certainly full, and punters heading for the trains shook their heads pityingly as I prowled around looking for a space. I went off to park elsewhere and came back half an hour later.

Alresford Station is done in Southern Railway or BR Southern Region green like, say, Swanage, but it did seem *exceptionally* green. On the platform there was even a green post box, in which deposited letters would be delivered by the railway – a member of the station staff (I assume) taking it from the green box and carrying it to any nearby red one. I had arrived on a day when a demonstration freight train would be running. The railway historian who'd told me that the Isle of Wight Steam Railway was a good line because 'they mow the banks', had also told me – in his pithy style – that the Watercress was 'good for freight'. You wouldn't *think* it would be good for freight, since the Watercress route is – and was – a country line, although not exactly a branch, because the Victorian Mid-Hants Railway, connecting Alton and Winchester, was part of a through route from London to Southampton. The line survived Beeching but was closed in 1973. The preservationists re-opened it along the 10 miles between Alresford and Alton in 1985, ten years after purchasing the line from BR. On the day of my visit, trains were not running all the way to Alton but terminating at the oddly named Medstead and Four Marks Station because of work going on to a bridge beyond there.

Historically, the line did carry quite a lot of freight: military equipment was taken between Aldershot and Southampton, and much watercress was carried, being despatched from Alresford to London in special ventilated hampers. Watercress is highly perishable, and its cultivation only became truly profitable with the advent of railways. It suited the local chalky soils, and Medstead and Four Marks Station is on a great lump of chalk, making it the highest station in south-east England. Train crews running through the station used to say they were going 'over the Alps'.

At Alresford, there was no sign of the demonstration freight train. It seemed to be quite elusive, and one volunteer from the line's Wagons Group, which operates the train, said darkly, 'Any slight problem with the passenger trains, and the freight's cancelled.' I eventually figured out that I would be able to pick it up at Medstead and Four Marks (henceforth Medstead) and possibly ride in its brake van, so I boarded the train to Medstead. This had a BR Standard Class 4 on the front, and an unusual BR type of BR Mark 1 buffet car: a griddle car, of which only eight were made. They were used mainly in Scotland from 1956 onwards and featured an anthracite stove for flash-frying of steaks. This one had curved banquette seats and small, stool-like Formica tables. It was at once both American and English – like a Wimpy Bar. I had a cup of tea, then went for a pee in the usual yellow Formica-and-wood Mark 1 WC. This had the thin tablet of yellow soap, which

I remembered from my boyhood railway journeys. When I returned to the buffet car, I congratulated the train guard on the authenticity of the soap. He frowned somewhat and went to consult a colleague. When he came back, he said, 'It's just cheap catering soap – we don't use it because it's particularly authentic.'

In the former goods shed at Medstead Station, whose goods *yard* closed in 1964, there is an interesting exhibition about the history of railway freight. Showing on a loop was the British Transport Films classic of 1957, *Fully Fitted Freight*, about a fast freight train from Bristol to Leeds, the speed enabled by the train being fitted with vacuum brakes. It's patronisingly addressed to a 'Mrs Smith', opening with an aerial shot of Bristol East marshalling yard and, 'Yes, goods wagons, Mrs Smith, just a few of the one million one hundred and twenty-four thousand eight hundred and twelve owned by British Railways.' The freight train featured in the film carried mixed goods, and there was a *Two Ronnies* 'Four Candles' flavour about the manifest, as described by the narrator: 'Seventy-two broom handles, one garden gate, twelve petrol engines, fifty-eight step ladders.' (This film is on YouTube and one of the comments reads: 'All made in Britain – what the f***'s gone wrong?')

Displayed on a wall was a blow-up photo of a page from the goods clerk's ledger of 1923. Among the figures in the 'Summary of Traffic' for Medstead in December 1923 are

Parcels forwarded: 389
Parcels received: 335
Milk churns forwarded: 1,109

Milk churns received: 10
Horses forwarded: 0
Other livestock forwarded: 4 cows
Coal forwarded: 0 tons
Coal received: 83 tons

Compared to other months, a much higher number of parcels was forwarded and received, and a higher amount of coal was received – a state of affairs presumably explained by Christmas and the cold weather, and I imagine the good clerk inscribing those figures in fingerless mittens. Other displays explained about the goods clerk's work. He had a *Handbook of Stations* so he knew what could be sent where: if a load weighed 5 tons and the destination station only had a 3-ton crane the load would have to be sent to the nearest station that *did* have a 5-ton crane. The clerk's bible was the *General Classification of Goods*, 'which assigned a wide variety of articles to approximately twenty categories'. I was reminded of a Victorian *Punch* cartoon showing an old lady travelling with a menagerie of pets. She is addressed by a ticket clerk: 'Station-master says, mum, as cats is "dogs" and so's parrots; but this here tortis is a insect, so there ain't no charge for it!'

The demonstration goods train had now clanked into the station, hauled by a sleek, black S15 Class engine of the LSWR. The station-master at Medstead, a very helpful and knowledgeable retired senior civil servant, watched its arrival with affectionate exasperation: 'Far too clean for a goods engine!' There were about a dozen wagons of various kinds, and all the details were correct for the BR era. The 'Freight Train' page on the line's website indulges in some quiet boasting:

> Correct painting, lettering and numbering of wagons and containers ... slight variations in the BR bauxite and grey liveries to indicate the effects of weathering ... authentic customer labels – i.e. BOCM (animal feeds); biscuits from Carr's of Newcastle ... evidence of straw packing from a previous load protruding from wagon doors ...

Even the chalk hieroglyphics on the wagon sides were accurate, although I think the legend 'Cowes' on the wagon from which a stuffed bull's head protruded was a joke.

A ride on the brake van *would* be possible, on payment of a fiver to the guard, the alternative designation of a brake van being 'guard's van'. The guard's little stove was not lit, presumably for health and safety reasons. This stove was about the size and shape of a traffic cone, and on an express goods train it would have glowed white hot because of the breeze. This being a preserved line, we were confined to 25 mph, but that felt much faster as I stood on the veranda of the van, and perhaps it *was* a little faster, since we were running downhill, and steam engines don't have speedometers. The drivers drove by instinct, whereas the goods guard was the strict timekeeper, hence the watch he would have been issued with in the old days. He could enjoy the ride, or perhaps have a kip next to his stove (and goods train guards vans were nicknamed 'guard's bedrooms'), but he couldn't sleep too long, since he would have had a couple of ledgers to keep updated, and he had to keep an eye on the vacuum brake gauge.

When we arrived back at Alresford I walked into the

town's greengrocery to buy some watercress, which I hadn't eaten for years – since when it has become (the greengrocer assured me) a 'superfood'.

So ended a highly educational day.

The Kent & East Sussex: a Colonel Stephens Railway

The Colonel

Colonel Holman Fred Stephens built rural railways on the cheap, exploiting the provisions of the Light Railway Act of 1896, which sets out what can and cannot be skimped in the creation of a railway, and on which Stephens was an expert.

Stephens was a Colonel in the Territorials and an engineer, trained at the Neasden workshop of the Metropolitan Railway – and he ought not to have been either of those things, given his aesthetic and bohemian background. He was named after the pre-Raphaelite painter, Holman Hunt, and his father, Frederic George Stephens, was a non-painting member of that brotherhood (he was an art critic).

Stephens was a hard-headed engineer and businessman, but there was an overlay of whimsy about him, and indeed his London club was the Eccentric. He ran a dozen lines from a house at 23 Salford Terrace, Headcorn, and was consulting engineer on many more. His lines did have platforms but not necessarily station buildings (or the station might be an old carriage), and they might not be fenced off from the adjacent fields. There might be no signals; crossing keepers were also unlikely: the fireman would get down from the footplate to open the gate.

Signals may be lacking, but then – to quote an old joke about branch lines – a single engine has yet to acquire the habit of running into itself.

Stephens' railways were standard-gauge, except his Rye and Camber Tramway in East Sussex (1895–1939) and the Ashover Light Railway (1925–50) in Derbyshire. The latter was about as far north in England as he roved. Of the Stephens lines that have survived in preservation, we could include the Ffestiniog and the Welsh Highland, which he was brought in to manage for a while in the 1920s, but the main ones are the Kent & East Sussex Railway and the East Kent Light Railway.

The original East Kent Railway was a standard-gauge network, mainly devoted to carrying Kentish coal to the port of Richborough. The fortunes of both the coal and the port waxed and waned quickly, and the network was dying from the late 1920s. In 1993, the East Kent Railway Society began operating trains over the southern part of the network between Shepherdswell and Eythorne. The preserved East Kent Railway memorialises the Stephens days in two signal-box museums, and its trains run through the only tunnel he ever built (he was forced into doing it by a local landlord). The ridiculously named Golgotha Tunnel is his most outrageous feat of ducking and diving: it has grand double-bore openings but is single-bore within, to save on excavation costs. But the East Kent has more on its mind than the cash-strapped Colonel. At Eythorne, for instance, there is a café in a BR General Utility Van designed to transport elephants; there is also the Walmer Model Railway Club, and its layout in an ex-LMS carriage. The line has mainly diesels, also some electric units, and is home to the Southern Electric Group and the Trolleybus Group. There is a strong electrical interest on the East Kent, and there has been talk of installing a third rail for electric preservation.

The main memorial to Stephens is the Kent & East Sussex, which runs for 11 miles from Tenterden to Bodiam in the Kentish Weald. It opened in 1905, as the Rother Valley Railway, running between Headcorn and Robertsbridge, thereby connecting two main lines of what was then the South Eastern & Chatham Railway. The Rother Valley Railway epitomised the light railways created under the 1896 provisions to assist farmers at a time of agricultural depression. On 3 June 1946, a poem called 'Farmer's Train' appeared in *Punch*. It was written by Hugh Bevan, with illustrations by Roland Emett, and the subtitle was 'Kent & East Sussex Line'. Here is the first verse:

Ever seen a railway train
Wheel deep in the wheat?
Poppies on the boiler dome:
wreaths of meadow sweet
twined about the driving wheel –
burnished brass and polished steel:
puffs of steam like woolly lambs,
on the line to Bodiam.

And here is the conclusion:

And Arcadia's just another station
On his twice-a-daily
pere-
grination!

Somewhere in the middle is the patronising line:

He sees real trains at Headcorn Halt

The Eccentric Club

What had been a grey morning turned sunny as I drove through Georgian Tenterden, and when I turned into the broiling gravelled car park of the Kent & East Sussex station, I knew I was going to enjoy the day, as did four people – two women and two men – of retirement age who climbed out of the adjacent car.

'It's lovely, isn't it?' said one of the women, surveying the little station, which had the peaceful aspect of a farmyard.

The other woman refined the remark. 'It's lovely and *quiet*,' she said.

Walking towards the station, the four became two couples. 'You know,' said the first woman, looking fondly at her husband, 'this reminds me of our honeymoon. Whatever direction we went off in, we always seemed to end up at a railway station.'

In the booking office, I bought a rover ticket, and a guide to the line, which featured a picture of a cricket match occurring with a tank engine hauling two green coaches just beyond the boundary. The legend read (somewhat tautologically) 'Travel back in time ... and escape the modern world.' Colonel Stephens always seemed to be urging people to do that on his railways, even back in the 1920s. An old poster in the booking hall advertised the line: 'Travel in safety across country, away from the dusty and crowded roads'; others demanded 'Support the local line!' The Rother Valley Railway was loss-making from 1924.

The railway escaped the grouping, but not nationalisation, and BR closed it in 1961. The preservation campaign was exceptionally slow. It was initiated in 1961 when the *Kent Messenger* published a letter from Tony Hocking, a pupil at Maidstone Grammar School. A preservation society was formed with the usual ambition of the early 1960s: to run a regular commuter service, but there were false starts in raising the money to buy the line from BR: the Rother Drainage Board and Kent River Authority believed the line would disturb the slumbers of the river, and so opposed the granting of the Light Railway Order. The line gradually reopened in its shorter form – about half the original length – from 1974. As preserved it is even sleepier than in the Colonel's day, since it is now separated by fields from its original junction termini, although there is a scheme – and in fact another preservation group, using the historical Rother Valley Railway name – to reinstate the 2 miles of line from Bodiam to the national network at Robertsbridge. (This new Rother Valley Railway has a visitor centre and shop at Robertsbridge: it is open on Sundays, but the RVR does not yet run regular trains.)

On the Kent & East Sussex, some blood-and-custard BR Mark 1s waited at the platform behind a simmering tank engine manufactured, apparently, in Sweden for Norwegian State Railways in 1919 – just the kind of maverick motive power the Colonel would have favoured (if he could have got it cheap enough).

I decided to let the train go without me. I'd catch the next one, even though I'd been warned quite severely by the ticket clerk (who looked like Miss Marple), 'It's a diesel.' I walked over to the elegant art deco café – which nonetheless has a corrugated iron roof. I then wandered over to the Colonel Stephens Museum. On the way, I was checked by the sight of a weird dusty vehicle, a conflation of a bus and some kind of rail unit. This was the Colonel's attempt to harness the power of the internal combustion engine that was perpetually undermining his railways. 'Based on a Model T Ford lorry,' says the line guide, 'this vehicle is capable of occasional operation, although the ride quality is best described as memorable.' The website of the Colonel Stephens Society is more direct. It says of his hybrid railmotors, which Stephens employed on the Shropshire & Montgomeryshire and the West Sussex Railway as well as the Kent & East Sussex, 'They were noisy, uncomfortable and reputedly drove people away from the railways they ran on.'

Entrance to the museum was free, which is just as well, since the attendant sitting outside in a garden chair seemed to be asleep.

'Afternoon?' I muttered experimentally.

'Afternoon,' he replied, without opening his eyes. He presided over what turned out to be one of the best small museums I have ever visited.

It was housed in one of those Second World War prefabrications, a Romney Hut – and so post-dates the Colonel – but it is roofed over with his favourite material, corrugated iron. Here is John Scott-Morgan, from *Railways of Arcadia:*

> On a typical station on a Colonel Stephens line, the platform was neatly kept, with an ashed surface and white-washed brick-faced edge. The building, made of corrugated iron or timber, was painted in a mellow

shade of buff-brown, its awning supports in a darker shade of brown. Inside the booking hall, with its timber-planked walls painted cream, upper part, and dark brown below, were benches along two walls in an L-shape. Opposite, there was a small pulley on the ticket office side, where there were racks of Edmondson cart tickets of all colours and types.

The Museum was astonishingly intimate, as though it had been curated by Stephens' own wife and children, but he had no wife and children. As the Colonel Stephens Society website notes,

> He died alone in 1931 ... in the Lord Warden Hotel, Dover, where he was resident. He never married and had no heirs. He left his estate of £30,000 (equivalent to about £1.4m today) to be equally divided among four of the staff at Tonbridge. This caused a great deal of dissension.

(This relative wealth is explained by the fact that the Colonel owned some of the railways of his empire, including the Kent & East Sussex.)

The museum re-created the atmosphere of a fussy, gloomy 1920s house, with floral wallpaper, numerous photographs and paintings on the walls, and cabinets loaded with memorabilia, such as Stephens' engraved cigarette case, his last unsmoked cigar, his masonic regalia and a silver mounted cane 'presented to Stephens on the extension of the Rother Valley Railway to Tenterden.'

Behind a glass screen sat the Colonel himself, in the form of a disturbingly lifelike waxwork. He sat at his own rolltop desk, surrounded by his own books and papers. On the desk were rubber stamps embossed with the names and addresses of the managers of his various railways. It was easy to imagine, from the ardency of his expression and the bristling look of his 'tache, that the Colonel had just banged one of these down on a memo reprehending the amount of money spent on (say) the unnecessary reupholstering of a third-class carriage. He could be a martinet, but also kind, as suggested by that bequest of his, or his habit of doling out cigars and gold sovereigns for meritorious performance. Light, but somehow ominous 1920s jazz was playing on a loop in the museum, gradually building tension: surely the Colonel would leap from his desk at any moment, seized by some new, cost-saving brainwave?

After the Museum, I boarded the diesel, which was a green Derby Lightweight multiple unit such as I had ridden on the Ecclesbourne Valley, and the ratio of window to carriage wall makes these almost like observation cars. I travelled in First, to watch the receding tracks through the windows of the empty rear driving cab. For this I was charged a supplement of £2 by a very genteel train guard: 'So sorry to trouble you ...'

The ride was so perfectly pastoral that it was quite featureless: sunshine, rolling meadows, occasional trees, sheep – no people. The single line of the railway disturbs the Rother Valley to about the same extent as the Rother river, which isn't much wider than the line: a dreaming green stream you'd think too lethargic to propel a fallen leaf, let alone create a valley. The topography, unchanged since the Colonel's time, allows the Kent & East Sussex to recreate the old Light Railway atmosphere without effort.

They could disrupt it by laying on amenities at all their stations, but in practice there are only things going on at the termini. At Wittersham Road there was a signal box and a passing loop (certainly no sign of any road), and I and about half a dozen other passengers stepped onto the platform to watch the steam train pass. When I proposed alighting from the DMU at the next stop, Northiam, and waiting for the steam train on its return journey, the guard and his colleague the station manager became quite agitated. I got the idea that nobody (except the station-master) had ever alighted at Northiam since preservation began. Didn't the line guide warn, 'Northiam is some way from Northiam village'?

At Bodiam – now, that is the place to get off. On the station there was a refreshment room and shop; there was a sort of picnic area on the old cattle dock. A few fields away, beyond the river, stood a – or *the* – castle. One of the intervening fields was a cricket pitch, and if a match had been in progress, the idyll would have been complete. The castle was a bonus to the Colonel, bringing tourist trade in summer on top of the farmers' traffic. He also benefited from the hops grown locally in his day, some of the Rother Valley being owned by the Guinness family. His trains carried hop-pickers to their encampments; he also ran 'Hop Pickers' Friends' specials.

On my way home, dicing with death on the M23, I thought about the Colonel, smoking his last cigar in the Lord Warden Hotel, which overlooks Dover harbour. He died after a long illness, so he would have had time to think about how he might be remembered. We don't know much about his motivation. In *Railways of Arcadia*,

John Scott-Morgan mentions the lack of primary research material on the Colonel: 'Much of the most useful archive material was destroyed by the staff of 23 Salford Terrace when the last of the active Stephens railways were nationalised.' He suggests that 'circumstantial evidence sets him in the scene of the Country House Movement and other late flowerings of Victorian Idealism which were destined not to survive beyond the 1930s.' Perhaps Stephens, for all his military rectitude, was an aesthete like his father. One thing is certain: he was losing his battle with the internal combustion engine, and he probably knew this. 'There should have been other Col. Stephenses running other groups of utility-built and cheaply run country railways,' writes David St John Thomas in *The Country Railway*: 'almost however badly maintained they would have been safer for local traffic than the parallel road.'

A Footnote: the Derwent Valley Light Railway

I want to mention another line that carried farmers' trains, but first, two disclaimers. The Derwent Valley Light Railway is not to be confused with the Derwent Valley Line on the national network. Secondly, it was not all that 'light'. In *The Derwent Valley Light Railway*, S. J. Reading observes: 'The Derwent Valley Light Railway was light in name but was surely heavy in presence. It was a standard-gauge railway from its inception, and it always carried heavy goods traffic despite its name and the rural locations of its stations.'

When I was a boy growing up in York, I had an elderly relative called Albert. He had a small farm near Wheldrake, which is about 7 miles south east of York, although it seemed like more back then. I would visit this farm, and leap about on the hay bales in the barn until Albert's wife warned me that I was in danger of landing on a pitchfork. Albert was less severe, and he would let me drive his tractor when I was about eleven years old.

When Albert spoke of 'the railway', he meant the Derwent Valley Railway, which amused my dad, who worked on BR, which had not deigned to take over the DVR.

The line had opened in 1912, running from York to Cliffe Common near Selby, where it connected to the wider world. It was at first mainly a passenger line, but passenger services ended in 1926, apart from occasional 'blackberry specials'. An article in the *Leeds Mercury* of 1928 described one such train calling at the village of Skipwith: 'The stationmaster's wife led one party down the line, and the stationmaster himself directed the others to the best areas for blackberries.' It also continued as a goods line for farmers, carrying potatoes, hay, grain, straw, oilcake, manure, timber and cattle. In the Second World War important government installations were created along the line (including for storage of petrol), because the line was so overgrown that it couldn't be seen by German bombers flying overhead.

In 1964, BR closed the Selby to Driffield line, so the junction at Cliffe Common became redundant and the DVR became a mere spur, and it closed in stages between 1965 and 1981. During that twilit time steam specials were run on the York end of the line from 1977 to 1979, in conjunction with the newly opened National Railway Museum at York. Did this make it a preserved railway? Not exactly, but today there is a preserved line using the original DVLR name and running over a short stretch of track (a three-minute trip) at Murton, where the Yorkshire Museum of Farming is located.

I think I remember seeing one of those 1970s steam specials on a drizzly day in south York, and this is the nearest I came to a genuine boyhood experience of steam. I certainly remember the York end of the line. The station was at Layerthorpe, where the DVLR connected to the Foss Islands branch of the York–Scarborough Line. This branch was surrounded by a lot of fading, mournful infrastructure including a gas works, a small power station and the 'destructor', where rubbish was burned, with black smoke drifting from a frightening, rusted chimney; there was also the vertiginous Leetham's Flour Mill (now very desirable flats) which overlooked the filthy (now clean) Foss river.

We seem to have come a long way from the notion of a 'farmer's train', and we move even further from that quaint notion when I mention that in 1984 the holding company of the original DVLR became Derwent London, a multi-million-pound property investment company.

Foreign Trains in Cambridgeshire: The Nene Valley Railway

Peterborough had been in the frame to host the National Railway Museum, but was beaten to it by another railway junction: York. However, the local authority had become interested in old railways and, in 1974, the Peterborough Development Corporation bought a stretch of what had been the Peterborough to Rugby line. This it leased to what was then called the Peterborough Railway Society which, in the mid-1970s, established the Nene Valley

Railway. Their stretch of line had been on an important cross-country route, and so the Nene Valley, like the Great Central and Gloucestershire Warwickshire, found itself on a generously proportioned railway.

One of the NVR's first acquisitions was a hulking, oil-fired Swedish steam locomotive (British steam locos being by now in rather short supply), and this helped determine the line's personality: it would use its scale, and its main line connection at Peterborough, to bring in and run European engines and carriages, which have a bigger loading gauge than British ones.

The NVR connects us to the golden age of the sleeping car, which occurred (with all due respect to the Caledonian sleeper) on the Continent from the 1890s to the 1970s, when the dark blue and gold carriages – sleepers and diners – of the Belgian Compagnie Internationale des Wagons-Lits traversed Europe. In the yard at Wansford Station stands Wagons-Lits sleeping car 3916, fading to sky blue under its tarpaulin while restoration funds slowly accrue. It was built in 1949 and is a veteran of the Nord Express. Close by stands another Wagons-Lits veteran undergoing restoration – dining car 2975. In the interior gloom you can make out luggage racks, an old suitcase, dusty pink lampshades. According to the NVR website, 'Number 2975 was built in 1927, and was used mainly in Switzerland, which is probably why it survived the war. It was brought to England for the filming of Stephen Poliakoff's television play of 1980, *Caught on a Train*, which is set on an Ostend-to-Vienna night train.'

Standing opposite this mouldering pair is a rake of pristine, unfaded midnight-blue carriages. Inside are cosy-looking compartments, and a 'Voiture-Salon-Bar' with beautiful marquetry and banquette seats. White enamel panels on the carriage sides identify them as having formed components of the second most famous Wagons-Lits service after the Orient Express:

Le Train Bleu.
(London) Calais–Paris
Nice–St Remo

'Actually, they're nothing to do with Wagons-Lits at all,' a volunteer at the NVR once told me. 'They're from the Belgian state railway in the 1930s. There's a clue *here*,' he said, pointing to a sign below the communication cord that read '*Noodsein*', meaning 'Alarm'. 'This was never a bar car, either,' he added. 'We converted it from ordinary seating.' The NVR runs Blue Train-themed dining evenings in these 'licensed Wagons-Lits-style carriages'. The fake bar car was used in the version of *Murder on the Orient Express* made in 2010 by ITV, starring David Suchet as Poirot. He convenes the suspects in the bar car in order to give his 'who done it' exposition. In the actual novel, Poirot convenes the meeting in the restaurant car; there never was a bar car on the Orient Express.

An orange and yellow Swedish diesel rail car called *Helga* is generally to be found shuttling up and down the 7-mile line. It is roomy and rounded at the ends like a tram or a boat, and it seems to sail serenely through the mild, Midland countryside. The toilet or '*Toalett*' is just a porcelain tube giving straight onto the tracks. When I first travelled on French trains in the 1970s, this was the normal arrangement. It was mesmerising to lift the lid and see the sleepers flickering by. Ostensibly this was primitive, not to say disgusting, but, being a Francophile, I immediately substituted other words like 'eminently practical' and 'unfussy'.

The NVR is home to the archive of the late George Behrend, an entertaining if louche character who was the number-one British expert on the Wagons Lits services, and possibly also the number-one consumer of its cordon bleu cooking and fine wines. It is also home to Phil Marshall, who is a director of the Nene Valley Railway Limited and lives in a flat on Wansford Station.

Phil is supervising the construction of an International Travelling Post Office Museum, to be opened at a new building at Ferry Meadows Station on the line. The museum will 'tell the story of delivering mail by rail', and will 'act as a bold illustration of the famous "Night Mail" poem by replicating the night-time working atmosphere of a busy station, where the visitors will be able to "step inside" the coaches that carried the night mail.' One of the Travelling Post Offices to be exhibited, TPO M30272, was on the train robbed in the Great Train Robbery. This being the Nene Valley, there will be a foreign element, courtesy of the International Railway Preservation Society, which is based at the railway. Punters will have 'a chance to glimpse the international transfer of mail as it makes its way across Europe, stored in the *fourgons* of another train of the night, the famous sleeping-car expresses of the Compagnie Internationale des Wagons-Lits ...'

Special Days

A Car Day

Road transport has been inimical to the railways in Britain since the 1920s and, as Richard Faulkner and Chris Austin point out in *Holding the Line*, the Beeching closures wouldn't have been politically possible were it not that car driving was seen as an enjoyable activity. It might seem odd, therefore, that classic and vintage cars are sometimes displayed on preserved railway sites, often on patches of land that were goods yards until road hauliers took the business away.

I suggested as much to some motorists exhibiting their cars in a car park adjacent to the Chinnor & Princes Risborough Railway, which preserves 4 miles of the old GWR Watlington branch in Buckinghamshire, which was closed to passengers in 1957. (A freight service continued until 1989, carrying coal to Chinnor cement works.) It was a lovely day in the Chilterns, sunbeams disporting themselves on the freshly waxed bonnets of the cars, some of which had leaflets inserted behind the windscreen wipers advertising the Oxford Bus Museum, the Marsworth Steam and Classic Vehicle Rally and other seemingly heretical causes.

My potentially provocative questions did not cause the slightest ruffle. 'Oh, well, it was the lorries that did the damage really,' said a man in charge of an Austin Allegro, as a GWR tank engine puffed away from Chinnor Station, just a hundred yards away.

He remained mild-mannered when I expressed surprise at seeing an Austin Allegro in company with E-types and MGs. 'We had one when I was a boy,' I added, 'and it was always breaking down.'

'That's right,' said the man. 'It would be. It was known as the "All aggro".'

Nearby, a couple were exhibiting a stately Alvis TC21. It had been owned in the 1950s by Sir Nicholas Sekers, managing director of West Cumberland Silk Mills Ltd. He had musical interests, being the chairman of the London Mozart Players and a governor of the Yehudi Menuhin School, and would often drive the Alvis down to London to attend concerts. I said, 'Just think of the first-class fares lost to the railways,' and the Alvis owners merely smiled. They saw no contradiction between an interest in cars and an interest in trains.

'It's all about British engineering,' the man said, and I guiltily recalled that the begetter of railway preservation, Tom Rolt, had not only driven a two-seater Alvis 12/50, but also ran a garage for a while and founded the Vintage Sports Car Club in 1934. Rolt's autobiography, *Landscape with Machines*, was well reviewed in *Motor Sport* in October 1971 by a petrolhead journalist whose querulous but respectful tone suggested he regarded Rolt as one of his peers:

> Interesting that Rolt describes the 22/90 RLSS Alfa Romeo as 'a not particularly good car' ... it was Col. G. M. Miles, not his brother Eric, who owned the Bugatti 'Black Bess' ... Shelsley Walsh [Hill Climb] is run by the Midland, not Midlands, AC, and surely the Mills 1907 Renault was bought by Marcus Chambers?

I moved on to a man exhibiting a couple of MGs and an E-Type Jaguar. 'I was a trainspotter,' he said. 'I copped most of the loco stock of the Western Region, but when steam was abolished, I lost interest.' Shortly afterwards, he bought his first vehicle, an A35 van. 'Then, when I was twenty-five, I got my first MG. Well, it was to pull the birds, wasn't it?'

Car people do tend to have more machismo than railway people, as I was reminded about an hour later, when the GWR Small Prairie-class tank engine had shuttled back along the line from Princes Risborough, and I boarded its train. Some blokes I recognised from the car show were sitting across the aisle from me.

'You ain't ever seen so many cameras in your bloody life,' one of them was saying, and he was not speaking of the half-dozen photographers who, a few minutes before, had been photographing the engine as it ran around its train. He referred instead to speed cameras on the A40. 'You used to be able to speed up between them – now it's average speed checks all the way.'

('Yes,' I thought priggishly, 'you're just going to have to obey the law for a change.')

The blokes did discuss the railway they were riding on, but in purely factual terms: 'You know about the extension? There's going to be a cross-platform interchange with Chiltern Trains at Princes Risborough.'

'I'd rather *drive* to Princes Risborough,' said one of his associates.

'Yes,' said a third. 'Lovely drive, that is – through Saunderton.'

('Poor old Saunderton,' I thought, still in priggish vein.)

They were right about the extension to Princes Risborough. It will be fully complete by the time this book is published. On my visit to the line, the C&PR trains were running into Princes Risborough station, but on a line one removed from the platform their trains will eventually use. So we couldn't get off or on at Princes Risborough, but we could contemplate the platform set aside for the C&PR, whose fixtures and fittings were already painted in old GWR light and dark stone. Previously C&PR trains had stopped at a buffer (and track loop) just outside the station. The removal of that buffer in February 2016, and the laying down of 10 metres of new rail to link to Princes Risborough was the culmination of twenty-five years of negotiation and planning.

Actually, that 'laying' should really be 're-laying': the C&PR will once again be a true branch, because a branch must have a connection to the main line. That said, I was told later in the day by the signalman at Chinnor that signalmen on the old branches would speak of the branch itself as the main line, to distinguish it from sidings off the branch. So it's all relative. The signalman had spontaneously offered me a tour of his box, and he began to explain his methods, but signalling is a blind spot to me. I was reminded of a sticker I'd seen on a railway DVD for sale at the Talyllyn Railway: 'Warning: contains description of block working.'

On the platform at Chinnor, a man came up to me holding a book containing a photograph of that last passenger train, of 1957, to run along the branch. He pointed to a young man. 'That's him,' he said, pointing to a much older version of the same man, who was sitting on a platform bench.

The Chinnor & Princes Risborough aspires to be 'a friendly GWR branch line' and it certainly is that. It is also a determined one. The Chinnor and Princes

Risborough Railway Association began re-opening the line in 1990. The station at Chinnor looks authentically Victorian, but the Association built it from scratch as a replica of the original, which had been demolished in the 1970s. The first train of the line's preserved era ran in 1991. Having re-connected to Princes Risborough, the Association plans to extend the other way, towards Aston Rowant. They will never re-connect to Watlington, though: the M40 is in the way. I have been critical here of motorists, but it should be noted that I used that road to get back home.

Christmas Days

All the preserved railways have Santa Specials, in which a volunteer dresses up as Father Christmas (on whom, unlike Thomas, there is happily no copyright) and dispenses presents. In theory these are jolly occasions as well as being lucrative, but in my mind there is a negative association between trains and Christmas. I'm thinking of two famous 'smashes'.

The first is fictional, occurring in Charles Dickens's ghost story, 'The Signalman'. The eponymous functionary – a sallow, neurotic figure – works in a box sunk in a cutting adjacent to a tunnel, where he is tormented by precognitive visions of a fatal crash. I associate the story with Christmas not so much because it first appeared in the 1866 Christmas number of *All the Year Round* magazine as because it was filmed by the BBC as their *Ghost Story for Christmas*, 1976, the film being shot on the Severn Valley Railway, using a fake signal box erected on the Kidderminster side of the Bewdley Tunnel.

The other smash involved real signalmen, and occurred

at Hawes Junction on the Settle & Carlisle line in 1910. At 5 a.m. on Christmas Eve a thunderstorm was raging on those moors, and a harassed signalman let the St Pancras to Glasgow sleeper run into two engines he'd left standing on the main line. Apparently he did not see the crash, but heard a faint whistle amid the storm, followed by a vague yet somehow decisive far-off rattle. A shepherd on a hill-side (a Christmassy-sounding person) reported seeing a single flash of flame. The signalman turned to a colleague in the box and said, 'Go to Bence [station-master at Hawes Junction, now called Garsdale, to which the Wensleydale Railway hopes to extend] and tell him I am afraid I have wrecked the Scotch Express.' Twelve people died.

The accident occurred at the highest point of the line, just beyond the north end of Moorcock Tunnel. The station immediately south of the accident site was (and is) Dent, whose station house is these days let out as a holiday home, and is very remote from the village called Dent. I stayed alone in the station one winter a few years ago, and on each successive evening (which I tended to spend drinking wine and reading about the hundreds of navvies who died in the creation of the Settle–Carlisle route) I became more aware of the blackened railway sleepers that had been planted upright in the ground as a snow barrier. They had become higgledy-piggledy over 150 years, like blackened teeth, or gravestones. They seemed to be advancing on the house but, as in the game Grandma's Footsteps, I could never catch them at it.

So my kind of Santa Special would have more in common with the Halloween 'Scary Specials' that some of the preserved lines now operate.

Father Christmas being the classic sole trader, it's ironic that Santa Specials were pioneered by a socialist.

On 18 December 1965, Bob Cryer, founder of the Keighley & Worth Valley Railway, stepped down from the cab of the line's first engine, the little Pug, dressed as Father Christmas. A small crowd awaited on the platform. He was not yet the Labour MP for Keighley, and the Keighley & Worth Valley Railway was not yet open for the carrying of passengers, which is why Santa/Cryer had to come to the public rather than the public coming to him. The accounts are vague as to what he did while dressed as Santa, but I do know from the KWVR publication, *Brontë Steam*, that Bob Cryer's mother Gladys ('a highly skilled tailoress') had made his Father Christmas coat. He apparently repeated the performance the following year, and *Brontë Steam* claims that Cryer 'pioneered a major slice of the heritage sector's business'.

The Santa Specials were only just getting going fifteen or so years ago, when I was regularly taking my children to preserved railways. Today, the Specials need to be booked in advance, and anyone signed up to preserved line mailing lists will be offered the chance to book somewhere around mid-June. It would seem odd – and possibly suspicious – for an unaccompanied adult male to book, but I did observe a Santa Special on the Leighton Buzzard Narrow Gauge Railway.

The booking office at Pages Park Station was swathed in tinsel. There was a touch of Christmas past, in that some of the shelves on which second-hand railway books were offered for sale from the collection of a volunteer who'd recently died. But it's Christmas, so I must be jolly. There was frost on the ground, which was quite Christmassy. A snowfall would have been atmospheric but possibly fatal to the enterprise, as on 10 December 2017, when heavy snowfalls across Britain led to many Santa Specials being cancelled (and the Severn Valley Railway had to hand back £35,000 in refunds to booked passengers). From a small table, refreshments were being dispensed: mince pies and orangeade for the children, deliberately weak punch – this was eleven in the morning – for adults. After Christmas, on this line and many others, the mince pies come to the fore with the trains called 'Mince Pie Specials'.

The children and their parents boarded the waiting train, as did I and my host, Terry, chairman of the line. As we pulled away I asked Terry (in a diplomatic whisper so as not to perplex any eavesdropping child), 'Who's Santa?'

'One of our firemen,' said Terry. Santa wouldn't be boarding the train on this trip, but would be waiting when the train returned to Pages Park. The engine was the most child-friendly on the railway, being called *Doll*, and liveried in a rich blue and lined out with red, therefore resembling a narrow-gauge Thomas. *Doll* was built by the firm of Andrew Barclay in 1919 and worked first at an ironstone quarry near Northampton. The owner of the quarry already had two other engines, and these were named after his daughters, Gertrude and Millicent. One of the two had a doll called Doll, and so the third engine was named after that.

Having disembarked from the train, the children, including a school party, were ushered into the room where Santa held court, from which they emerged beaming and carrying big boxes in blue carrier bags. I asked Terry what was in the boxes, and he wasn't sure.

The presents – graded according to age – had been bought as a job lot from 'a Christmas present company'.

Another volunteer (also not privy to the contents of the boxes) said, 'I do know this: we had a store room four metres square, and it was absolutely full of the presents.'

The Santa Specials had been a great success, as usual. 'We had five thousand people for them last year,' said Terry, 'and we've had nearly that many already this year. That compares to thirteen thousand we had in our main summer season.'

Before I left the railway, I asked one of the children, a boy of seven, if he had enjoyed the day.

He nodded.

And would he be coming back?

Another nod, but his father chipped in with, 'We're regulars here.'

'What did you enjoy most about the day?' I asked the boy.

'This,' he said, holding up the still-mysterious big box.

*A Steam Driving Day**

In Cornwall, where the 125 mph diesels from Paddington proceed at 25 mph so as not to damage viaducts built by Isambard Kingdom Brunel, lies Bodmin Parkway station, formerly (and more honestly) known as Bodmin Road. Here, at the right time of year and day, you can transfer to the highly successful Bodmin & Wenford Railway, over on Platform Three. From there a steam train will take you along the 4 miles of line (built in 1887, closed in 1967,

* This section is a slightly adapted version of an article originally published in *Vanity Fair On Route* (2017) © Conde Nast Publications Ltd.

reopened in 1990) to Bodmin itself, and the station self-effacingly called Bodmin General.

Bodmin General is HQ of the B&W Railway, and its mid-point, lying as it does in between the two B&W termini at Bodmin Parkway to the east and a lonely spot called Boscarne Junction to the west. (And the railway aims to project beyond Boscarne to Wadebridge).

The day-long steam driving course was scheduled to begin at 9 a.m. It costs £395, whereas a 'heritage diesel' driving day would have cost £280.

At 8.30, I arrived at Bodmin General to discover a tank engine coming to a boil in an engine shed whose interior was made to seem dark by the brightness of the morning. The engine, 6435, was one of the 6400 class built by the GWR in the 1930s for branch-line work. A lean, twinkly-eyed man was standing near the loco with a proprietorial air, and he introduced himself as Jimmy James, one of our two instructors for the day. Jimmy was seventy years old, a retired army major: 'As a boy, my only ambition in life was to fire or drive steam trains, but my parents had other ideas.' In other words, he was too middle-class for the profession of railwayman, but certainly not for volunteering on a preserved railway. In my experience these are socially, if not ethnically, diverse. They remind me of Waterloo Station as described by P. G. Wodehouse: 'The society tends to be mixed.'

Jimmy led me over to the station café, where I met my two fellow students. One was an electronic engineer, the other a boat builder, with experience of being 'third man on a traction engine'. 'But we get all sorts,' said Jimmy: 'lawyers, teachers, doctors ...'

'Any women?' I wondered.

'The occasional woman, yes. We once had an eighty-year-old retired nurse.'

Our second instructor, Dave Holford, now appeared, holding a mug of tea, and looking dapper in two-piece denim overalls. He mentioned that he was a paramedic in real life, which I found reassuring after listening to his safety briefing. We were all here for the love of steam engines but, as Dave reminded us, 'They'll kill you if they get the chance.'

For our first exercise, we would all have a go at driving halfway to Boscarne Junction. I unscrewed the hand brake of 6435: a typically elemental activity, in that it involved turning a great steel wheel twenty times in an anti-clockwise direction. The initial stiffness and the sudden stop both present wrist-breaking opportunities. The steam for propulsion is applied by the satisfyingly-named 'regulator' – a steel stick mounted horizontally. You only have to lift the end of it a couple of inches to send steam into the cylinders, where it moves the pistons that drive the wheels. The paradox is that, in order to drive a machine weighing at least 30 tons, you need a light touch. Here is Flann O'Brien, writing as Myles na Gopaleen, in a typically fantastical piece entitled 'For Steam Men', written for the *Irish Times* and collected in *The Best of Myles*:

> I should not like it to be taken from my remarks a few weeks ago that I would not like to go down in railway history as a 'full regulator man'. In my railway days you would not always find me with the lever pulled right up; still more rarely would you find me yielding

to the temptation to work on the 'first port' and cut off late. Sometimes, when conditions suited, you would find me blowing down to reduce priming, but never when the design left me open to the danger of having my valve jammed open. And none knew better than I when to shut off my cylinder oil and feed when drifting.

The first few movements of the pistons make a compelling noise. Think of a man walking through powdery snow: one lurching crunch after another, gradually increasing in speed until you have what is usually called 'chuffing'.

We were running 'light engine' (without carriages) to Boscarne. But at 45 tons, 6435 is not *particularly* light, and you are very conscious of the weight, because the two lines of the B&WR ascend towards Bodmin General at a gradient of about 1-in-34. A car will go up a 1-in-3 with no trouble. A steam engine will go up a 1-in-34 *with* trouble, especially if the fire is not burning fiercely enough to generate sufficient steam. But going down is rather frighteningly easy, and Dave kept saying 'Let it roll, let it roll.' An occasional touch on the brake was all that was needed, and I wanted to touch it more frequently than Dave requested, such was the clanking, rattling and jolting going on around me, with nothing that was not filthy or hot to grab onto. Engines on preserved steam lines are restricted to 25 mph, but 20 mph on a tank engine going downhill feels like 100 mph. Dave said he rarely hits top speed. 'You run out of courage before you get there.'

After we'd all been to Boscarne and back, we repaired

to the café for tea and all we could eat of the chocolate biscuits. The next run was to Bodmin Parkway, but this time a dirty, ridged pipe, somewhat resembling an elephant's trunk, connected 6435 to three carriages. Within this pipe, a vacuum could be formed or destroyed by swivelling the same lever that controls the steam brake of the engine when 'running light'. The vacuum holds the carriage brakes *off;* when the vacuum is destroyed, the brakes come on. The system ensures that, should an accident cause the vacuum pipe (and therefore the vacuum) to be broken, the train will stop.

This might be a fail-safe arrangement of brilliance, but the vacuum brake confused me. I kept creating rather than destroying the vacuum, then overcompensating the other way. This resulted in a jerky ride for the four people in the carriages (our train guard and three friends of the electronic engineer), and Jimmy – who had urged me to imagine I was towing a restaurant car full of genteel diners – would satirically observe, 'Nose in the soup again!'

That was going downhill to Parkway. Half an hour later, as we started the ascent, rain began to fall, and I was admiring the drama of the lowering skies when Dave mentioned the weather's true significance. 'The wheels might slip; you'll know about it if it happens.' It *did* happen. Imagine one of those LP records of a steam locomotive chuffing away; now imagine a scratch on the record causing the needle to jump. I felt no actual backward movement, but it was as though the engine's heart skipped a beat, and so did mine. But Dave intervened immediately, telling me to 'Shut off steam!' (An engine can shake itself to pieces if it begins running on the spot for any length of time.) And Jimmy was on the case from his side, activating the lever that dropped sand onto the rails, so giving us the friction we needed to proceed.

After lunch – a delicious and authentic hot Cornish pasty – I had a go at firing to Bodmin Parkway. In steam days, an aspirant driver might spend twenty years firing before graduating to driving, which would have been a very satisfying promotion, since the driver was the footplate boss: he got to remain fairly clean; he was better paid and had more status. It was his hand the grateful holidaymakers wanted to shake when the train had taken them to Blackpool or Penzance.

But it is the fireman who makes the engine go. He must ensure the fire is hot enough to maintain the required steam pressure. But if the pressure is too great, the safety valve on top of the boiler lifts and the engine begins to scream, while shooting steam into the sky, where it condenses and rains down on all those in the vicinity. This is highly embarrassing, as when you're cooking for a dinner party and the smoke alarm goes off.

The fireman must watch not only the steam pressure gauge, but the gauge that shows how much water is in the boiler. If he fails to keep this topped up, the boiler will explode, and he and the driver will die. It is also the fireman's job to couple and un-couple the carriages, which must be done at every terminus, when the engine 'runs round' its train before attaching to the other end. The fireman steps onto the tracks between the engine and the first carriage – which is called 'going in' – and everybody involved with the train must know he is there. It is then a matter of humping the great shackle onto the monstrous hook, before connecting up the two ends of the vacuum tube. I 'went in' once with Jimmy, but he took one look at my leather gloves (chosen for aesthetic effect rather than practicality) and shook his head: 'You'll ruin those,' he said. 'Better leave it to me.'

The fire intimidated – but also fascinated – me: like the sun. It's hard to look at, but you must constantly appraise its status. And the coal was not the sort you could put on using tongs. Some of the lumps were bigger than my head ... and it takes years to learn how to flick them off the end of the shovel to the front of the firebox, so I kept having to hand the shovel to Jimmy. But he was kind about my firing ('Very solid'), which I suspect is C-plus.

As I took the train back to London that evening, my fellow passengers all looked unnaturally clean, and the technology they were concerned with – their mobile phones – seemed absurdly fiddly and minute. Over the next couple of days, the world continued to seem plastic-y and unauthentic. I missed the soft, warm embrace of the steam, and the reassuring (in retrospect) glow of the fire. In the course of writing a number of books about railways, I have interviewed a number of ex-steam train drivers who, without exception, loved the job. My day on 6435 gave me some idea why.

Thomas Days

My youngest son is famous (in our family) for having said of an arriving Tube train when he was three years old or so: 'It has no face.' I presume the reason he expected it to have a face was that I was just then force-feeding him the Thomas the Tank Engine stories, in which engines do have faces. (When the Reverend Awdry first thought up the stories, as a means of entertaining his son, he drew the engines himself; it was easier to draw them head-on, and the crude circle of the smokebox demanded a face.) I would read Awdry's stories to both my sons at bedtime, and they would watch videos of the television versions animated by Brett Allcroft and narrated (the early ones, at any rate) by Ringo Starr.

In 1990, an event was held at the National Railway Museum to celebrate the Thomas books, which 'have played an enormous part in arousing children's interest in railways'. At the same time, Charles Jennings wrote an article for the *Evening Standard* entitled, 'All steamed up over Thomas', in which he described the Reverend Awdry's prose style as 'at best workmanlike, and at worst narcotic'. He has a point, but I think I was always hypnotised by the brightly coloured illustrations, and Jennings concedes, 'for all their awkwardness they manage to convey the sense of a bright, clean unvandalised world, located somewhere in the mythical 1950s.' The books are postcard-sized (an early anxiety of the publishers was that they were small enough to be slipped into a pocket and stolen), and they are like postcards from a sunnier world.

My younger son first saw a 'live' Thomas – that is, a tank engine with a plastic Thomas face strapped over the smoke box – a couple of years after he'd commented on the facelessness of Tube trains, so I was interested in his reaction.

I said, 'What do you think?' and he said, 'It's pasty', a word I did not know he knew. There was no question but that he was speaking of the pastiness of an artificial face rather than anything he thought real. I did once see a very small child start crying when he saw the Thomas face on the front of a tank engine at the North Yorkshire Moors Railway, and I think the child was scared because he thought this really was a living engine and therefore something supernatural.

Thomas Days are the biggest money-spinners after Santa Specials, but not as many lines stage them as used to. Twenty years ago, thirty lines might have held Thomas Days every year. Now it's down to a dozen or so, and they are the bigger players. Thomas is a more expensive commodity these days. In the past, a Thomas Day might have involved a fairly fleeting station run-through by a tank engine with a Thomas face strapped onto the front. Since then, successive Thomas franchise owners (the current one is HIT Entertainment) have required the railways to mount a bigger production; the royalty payable to them (the percentage of the takings) has also increased. Thomas Days really must last a day. They might involve some of the attractions local to the railway (those on the South Devon Railway involve a visit to the rare breeds farm at Totnes) as well as re-enactments of scenes from the books: for example, the Fat Controller supervising the watering of Thomas in appropriately bumptious manner.

In the summer of 2001, a volunteer on the South Devon, John Keohane, came up hard against one of the strictures. Mr Keohane is the South Devon Railway's 'registered Fat Controller'. He had achieved that status in spite of having a beard, whereas Sir Topham Hat (as the Fat Controller is more respectfully known) is baby-faced. As widely reported at the time, Mr Keohane was asked to shave off his beard. It was also reported that Mr Keohane was 'a Beefeater at the Tower of London' and that 'all Beefeaters must have beards.' Two clarifications are in order: first, Mr Keohane was not merely a Beefeater, but Chief Yeoman Warder at the Tower; second, it is not a formal requirement that Beefeaters have beards; it is just a tradition. So Mr Keohane shaved off his beard.

Then there is the up-front cost of obtaining a 'Thomas'. I was about to say that Thomas the Tank Engine is 'based on' an engine of the E2 Class, designed by the superbly named Lawson Billinton: ten were built by the London Brighton & South Coast Railway between 1910 and 1913. But given the Reverend Awdry's pedagogic approach, it would be better to say that Thomas *is* an E2 ... and there are no E2s left. The last one was scrapped in 1963.

So various sorts of tank engines are mocked up as Thomas, then leased out between the railways. In 1994, engineers at the Mid-Hants converted an Austerity 0–6–0 saddle tank into a Thomas, which required quite major surgery, since Thomas has side tanks. At one point the South Devon had its own mock Thomas, but this was a static engine, not steamable, although it did talk, or rather a man in some remote location talked, and his voice was relayed through a speaker implanted in the mocked-up engine.

An official of one small preserved railway told me, 'We can no longer afford Thomas. We have Paddington Days instead, which are much cheaper.' Paddington has strong railway credentials, reinforced by the two feature films, both of which contain railway scenes. He is named after a station and was created by a man – Michael Bond – who was a genuine rail enthusiast. But Paddington himself does not run on rails, so the person dressed up as him has to be strategically located. The impersonator is always placed at the terminus so the children have to travel along the line to meet him. (I think Julia Donaldson's creation, the Gruffalo, also deserves to get some preserved-railway action, on the strength of her response to the following question, asked by the *Guardian* in August 2011: 'If you could bring something extinct back to life, what would you choose?' She replied, 'The railways axed by Beeching.')

Other 'characters' in the Railway Series are easier to present. Almost any old guard's van will do for Toad, for example. And as Awdry himself stated at the start of the book, *Oliver the Western Engine*: 'Readers may like to know that "Olivers" and "Ducks" still work on the Dart Valley Railway in Devonshire' (which is what the South Devon Railway was called until 1992).

The mixed traffic engine called Henry – 'bigger than James, but smaller than Gordon' – is sometimes represented. He is the most interesting character in the series, and the most problematic. Several of the controversies about Awdry's work swirl around Henry, and these might keep some right-on parents away from Thomas Days. An article in the *New Yorker* by a writer called Jia Tolentino spoke of 'the repressive, authoritarian soul of *Thomas the Tank Engine and Friends* ... Wilbert Awdry, who created Thomas the Tank Engine, disliked change, venerated order, and craved the administration of punishment.'

The article is about the TV versions of the stories, called *Thomas & Friends* in the UK and, creepily, *Shining Time Station* in the U.S. Tolentino focusses on the version of one of the stories appearing in the first book in the Railway Series, *The Three Railway Engines*. Henry is not one of the three, but he has a starring role in the third of the four stories, 'The Sad Story of Henry'. In this, he takes refuge from the rain in a tunnel and refuses to come out; so he is bricked up in the tunnel: 'They took up the old rails, built a wall in front of him, and cut a new tunnel.'

On the final page of the book version Henry is told, 'We shall leave you there for always and always and always', and the narration ends, '... I think he deserved it, don't you?' Tolentino concedes that this last line is tweaked in the American TV version to suggest some possibility of parole.

Tolentino rounds up some of the usual criticisms of the stories. These were summed up in a *Guardian* article of 1991 by Mike Jarrett: 'women and workers both know their place. The coaches, Annie and Clarabel, are content to tag along with Thomas, going wherever he takes them and cooing at his manly feats and superior wisdom. The workers, that grey, grimy mob of troublesome trucks, occasionally stage mindless rebellions until some rough shunting puts them back where they belong.'

There was a fit of criticism of the stories at about this time. John Westerby, Chief Education Officer of Dudley, wrote a long report (leaked) questioning Awdry's politics: 'the engines are male, while the female coaches were merely pulled and shunted ... The message, even if unintentional, conveyed to young minds is clear: men lead, women follow.'(It is always the unintentional messages the authors have to watch out for.) A letter to the *Guardian* suggested that the books were also homophobic. 'When the fastidious Henry refuses to come out of the tunnel because the rain will spoil his nice red [sic] paint, he is ostracised by all. Is the tunnel here a metaphor for the closet?'

Henry's nice paint was actually green. According to Awdry's biographer, Brian Sibley, Awdry had not at first intended to release Henry from the tunnel, but he did eventually decide to conclude that first book with a story in which he is freed and painted blue. This was problematic because Gordon was also blue, although he could be differentiated by his greater size and his square buffers. But when Henry reappeared in the fourth book of the series, *Tank Engine Thomas Again*, the illustrator, C. Reginald Dalby, gave the newly-blue Henry square buffers as well. According to Brian Sibley, this caused 'a furore of indignation' among young readers. That was in 1949; I doubt that today's young readers would be so attuned to a railway detail.

The main Henry book is number 6 in the Series, *Henry the Green Engine*, and this is missing from a set I bought in an Oxfam bookshop in Hampstead. It had possibly been jettisoned by some Hampstead liberal, because in the first edition of this book, which appeared in 1951, Henry (whose buffers had reverted to being round) puffs smoke over some boys throwing stones from a bridge, so they end up 'as black as —', the 'N' word being employed. This was pointed out in the *Sunday Times* in 1972. Awdry's initial response was peevish: 'I don't know what the alternative will be,' he was quoted as saying. 'If you say "as black as chimney sweeps or miners" you have their unions up in arms.' He did amend the passage so that it read 'black as soot', but the 'N' word persisted in some editions, and a woman who had adopted a West Indian baby blacked it out when she found it in a library in 1976. She wrote to Awdry, who apologised graciously to her.

(It ought to be mentioned that the Thomas stable is becoming more diverse, at least on TV. The new series featured a more multi-cultural and gender balanced 'steam team', and new characters include female engines Hong Mei from China and Ashima from India.)

At present, though, it is Thomas himself who remains a synonym for 'steam locomotive' among so many young children. His creator, Reverend Awdry, emerges pretty creditably from Sibley's biography. He was, like Tom Rolt, born into a social class higher than that associated with railways, so there seems something magnanimous about his interest in them. His father, also a clergyman, had a railway in the garden of the vicarage at Ampfield in Hampshire. It was landscaped to the extent that the train would disappear entirely from sight when it ran through a cutting.

The young Awdry would walk to a platelayers' hut made of sleepers at a spot near Ampfield on the Wessex main line. He would talk to the platelayers as they smoked their pipes, experiencing 'an atmosphere almost of a fairy-tale quality'. In 1981, he invited Ringo Starr, newly installed as narrator of the televised stories, to his house at Stroud, where he demonstrated his model railway. 'Tell me when you're bored,' he said.

'Not yet,' Ringo gamely replied.

Using a radio control, Awdry remotely detached trucks from lines of goods wagons.

'Cool!' said Ringo.

A journalist covering this summit meeting of the two cultural icons for the *Mail on Sunday* wrote, 'Mr Awdry was reminiscing for our benefit in that gentle spellbinding way of all good storytellers, when Ringo said he was sorry, but he had to go. His wife roused herself from a state of almost catatonic boredom and they departed ... Mr Awdry bade them a courteous, abstracted farewell.' The Reverend Awdry replied to all readers' letters – usually from boys with names like Peter, Richard, Giles;

indeed, he retired prematurely from the church in order to do so, and he was generous with charitable donations.

He certainly did plenty of favours to the preserved railways. We have discussed his boosting of the Bluebell in *Stepney the 'Bluebell' Engine*. It was on the Bluebell, in 1979, that Awdry met Britt Allcroft, who first put Thomas on TV. She'd been making a film for the Central Office of Information about the resurgence of steam, and reading Awdry's book, *A Guide to the Steam Railways of Great Britain*. Over dinner on Bluebell premises, she told him that his Railway Series (whose books had sold eight million copies at that point) deserved to be brought to life.

As the later stories become enmeshed with the real histories of preserved railways they deteriorate, becoming full of arch references and in-jokes, but perhaps this was all in a good cause. The penultimate book in the series, *Duke the Lost Engine*, opens with a plug for the Ffestiniog Railway and the Ravenglass & Eskdale. That story features the Thin Controller, who was based on Tom Rolt, and the Fat Clergyman, who is based on Awdry's friend, the Revd Teddy Boston who – on two Saturdays a month – opened to the public his 2-foot-gauge Cadeby Light Railway located in the grounds of the rectory of All Saints church at Cadeby in Leicestershire, of which Boston was the incumbent. Boston died in 1986, but his widow kept the line open until 2005. The Moseley Railway Trust runs some of his stock, including an engine called *Pixie* (ex-of a quarry at Cranford), in Apedale Community Park, Newcastle-Under-Lyme.

Boston was a jolly, roly-poly figure of a man. A stained-glass window in his church showed a narrow-gauge engine and the words, 'We Acknowledge Thee To Be The Lord.' He was a sometime chair of the Narrow-Gauge Railway Society. He also had a large double-O-gauge model railway and played clavichord and bassoon; and he wrote Biblical parodies. Here is the beginning of 'The First Account of the Zealots called REGENESIS, which being THE PARABLE OF THE PUFFING PIXIE.' It begins:

1. In the beginning Bagnall created the type 2090 and called it PIXIE.
2. And the PIXIE was used and cast aside; and darkness was upon the footplate. But the spirit of zeal did move nigh.
3. And the spirit did see that the PIXIE was good and did place protection upon it; and it was conveyed to the haven called Rectory.
4. Thus it came to pass that about the hour of noon, on the day before that of the Sun of the Palm, there did congregate several zealots at the haven called Rectory; which was in the land by the Cade.

Ale Train

One day in early 2017, I read, in the Campaign for Real Ale's free magazine *London Drinker* (available in all good pubs), 'EPPING–ONGAR RAILWAY STRENGTHENS TIES WITH CAMRA FOR REAL ALE TRAINS'. The piece gave news of a Spring Beer Festival to be held on the Epping & Ongar preserved railway in East London.

It was raining heavily as I set off for my local Tube station. Soon I was rattling east on the Central Line which, after it comes above ground at Stratford, seemed to be taking me not only further and further east, but also back in time. The stations became more countrified, with such antiquated grace notes as valanced canopies held up by barley-sugar iron stanchions, and public lavatories.

Many of these stations had been built by the Great Eastern Railway, and were only claimed by the Underground in 1946 or so when, as part of the New Works programme, the Central Line was extended east from Liverpool Street. The extension reached Ongar, but the final stretch, between Epping and Ongar (with stations called North Weald and Blake Hall in between), always operated as a country branch line of the Underground, and you always had to change at Epping to access the wider network. At first, moreover, the branch was operated by steam engines (so much for 'New Works'). There weren't enough people between Epping and Ongar to justify electrification, and further development was inhibited by the imposition of the Green Belt.

John Betjeman was a fan of Epping–Ongar, and said that when he retired he'd like to be the station-master at Blake Hall, which seems a bit condescending towards the actual holder of that job, but then again in 1981, when Blake Hall closed, it was being used by an average of six people a day. But we are getting ahead of ourselves.

In 1957, Epping–Ongar was fitted with 'light electrification', which sounds pleasant enough, and was sufficient for Tube trains of three or four carriages in length to work the line. By the early 1990s the shuttle was being used by around eighty people a day and it was said – to me, by an inebriated man in a pub in Leytonstone – that the line was only being kept open so that, in the event of a nuclear attack, members of the Cabinet could be evacuated by rail to the nuclear bunker at Doddinghurst. The argument against this theory is that surely no government would entrust its fate to the Central Line. Then again, the Ministry of Defence did de-commission the bunker at about the same time London Underground closed the Epping–Ongar line, which happened in 1994. In 1998 the line was sold to a private company, which operated a peak-hour commuter service for a while. (The nuclear bunker, meanwhile, was acquired by a local farmer, who advertised it widely as 'The Secret Nuclear Bunker!')

The line was sold to a company called Pilot Developments in 1998. Their intention was to run a regular passenger service, but this did not materialise. In 2000, the Epping–Ongar Railway Volunteer Society was formed. In 2004 the Society began running trains along the line on Sundays. In 2008, the line was acquired by Mr Roger Wright. It is part of the mythology of the EOR that when Mr Wright, who made his money running an East London bus company, decided to cash in his chips, he was torn between roving the world on cruise ships or buying a rusty old railway. The line was then closed until 2012, during which time infrastructure improvements were made to allow Epping–Ongar to be a fully functioning preserved railway open on weekends, Bank Holidays and some summer weekdays.

At Epping Station, I boarded the bus laid on to link to the railway. You would think you could just change onto the preserved line by crossing to another platform at Epping Station, as in the days when Epping–Ongar was part of the Underground, and the preserved line of today does run to within a hundred yards of Epping Tube station, but that point is a buffer stop in Epping Forest, and you can't alight or board there. The nearest boarding point for the EOR is at North Weald Station, hence the need for the bus, which was, naturally, a heritage bus: a double decker, not a Routemaster but a Regent Type, or RT, which was *the* London bus of the 1950s – and this was a green one, having once been used on the Green Line country services of London Transport. In spite of the need to board this bus, the Epping–Ongar is (the narrow-gauge Hampton & Kempton Waterworks Railway aside) the closest preserved line to central London, allowing it to boast of providing 'the only Real Ale train you can reach using an Oyster Card'.

The bus had a conductor – an older gent, let's be frank – and when, as he approached me with his antiquated Gibson ticket machine, I asked him the fare, I subconsciously expected him to say something like 'One and six'. When he said, 'That'll be eighteen pounds, please,' I was a bit shocked. 'That's for everything, you know,' he explained – 'all day long.' The cost would be not only my bus fare, but also my ticket for the trains, including the Ale Train, although not the actual ale.

It was midday as we pulled away from Epping Station, and still raining, but the weather made the green and brown bus interior seem cosy, and reinforced the Englishness of Epping's suburbia. Not everyone was for the railway, and when people got off further up the road in Epping the conductor, with his instinctive talent for the old-fashioned, would say, 'Mind how you go.'

Other vintage buses had been parked up near North Weald Station. The wagons had been circled, as it were, and the place was abuzz with London-centric bus and rail enthusiasm, which tends to be less romantic than you might expect. A marquee erected on the station forecourt was devoted to the sale of memorabilia related to diesel and electric multiple units, and other fairly recent phenomena of the slam-door commuting life. The Network Southeast Society had a stall in the marquee, manned (that is the operative word) by some of the most articulate and passionate enthusiasts I have met; but I found their interest in that operational subdivision – or 'business sector' – of late-period BR slightly baffling. To me, 'Network Southeast' lacks the resonance of 'Southern Region'. I associate it with bleak red, blue and white livery and signage, applied to undermanned stations in which CCTV cameras had taken over from people, hence (on platforms) mirrors allowing a driver to see along the length of his train without the aid of a guard; or automatic ticket machines; or vaguely Orwellian signs saying things like, 'We care about your safety. Please do not join or alight from a moving train.'

The actual station of North Weald – decked out in London & North Eastern Railway green (the LNER having inherited it from the Great Eastern in 1923) and the ticket hall with a good coal fire burning in the grate – accorded more closely to my own idea of railway romance. I boarded the next available train, which comprised an assortment of Mark 1 and 2 carriages (plus a diesel multiple unit carriage not accessible from any of the others), hauled by a green Class 37 diesel. This train became a Real Ale Train, as it were, gradually. Plenty of drinking on board was going on during the early afternoon, but the official Real Ale regime was not scheduled to begin until five o'clock.

I settled down to an afternoon of shuttling along the

line between Ongar and the line's most westerly point in the green glade in the woods, where the sound of the rain rattling on the leaves was audible even through the closed windows of the train. Every time I passed Blake Hall Station, I looked for any sign of the owner of the railway, Roger Wright, who apparently fulfils the Betjemanesque fantasy of living at the station. After the station was closed in 1981, some Underground driver naughtily (or kindly) continued to stop there to allow the small trickle of local users to board or alight, so the Central Line management removed the platform, which has been reinstated by the preservationists.

Sometimes, in my shuttling up and down, I alighted at Ongar, which the EOR people like to say is the only ex-Great Eastern Railway station in its original colours. These had been unearthed by 'extensive research', yielding a bathetic conclusion: three shades of brown. At Ongar there is a buffet, a gift shop, a local history-themed exhibition called the 'Penny Salon micro gallery', and a roofless Gents. A sign reads:

> We apologise that this toilet has no roof at present. It blew off during a recent gale. Please be careful as floor may be slippery.

There was a footnote:

> This toilet actually had no roof when it was built in Victorian times but we are not that mean and it will be replaced shortly.

After my station visits, I would adjourn to the same

plush, pink-upholstered compartment, which I always had to myself. At about four o'clock, during one of the woodland stops, a tipsy group came clattering along the corridor.

'Choo choo!' shrilled a man amongst them.

One of the women said, 'Oh, look at all these little rooms!' They had now arrived alongside my own peaceful 'little room'.

'We can ask if we can go in there,' said one of the women. 'There's only one man.'

'No, we can't,' said the other woman. 'He's writing.'

'Don't be silly,' said a man. 'You don't have to ask.'

But they did ask. 'Is it all right if we sit here?' said the woman.

'There's twenty-eight of us,' added the man, which was a joke. There were half a dozen in their group and only five available seats (after I had collected up the sprawl of railway books I'd been reading), so one man remained loitering (and drinking) in the corridor. The five who entered the compartment were drinking as well, and some of them were also eating.

'Pork pie?' one of the men asked me. 'Pickled egg?'

Another of the men was reading some EOR literature. '"North Weald"', he said, '"is 340 feet above sea level. So when it was part of the Underground it was the highest point *on* the Underground." ... I bet *he* knew that,' he said, indicating me.

When, at 17.10, we pulled away from North Weald we had officially become the Real Ale train. A converted brake van had been attached, and this, with the aid of a trestle table and some old bus seats, had been converted into the Pig & Whistle Bar. By 17.12, the Bar was packed.

There were three barmen, filling glasses direct from the barrels. I'm not usually a Real Ale drinker (I prefer white wine), and I think the barman who served me was disappointed when I cut him off midway through his descriptive speech about one of the beers by saying, 'Sounds fine, I'll have a pint.'

One of the barmen, Alan Perryman, was also the organiser. Alan is on the management committee of the railway and a member of the East London branch of CAMRA. He began running beer festivals at university. 'A serious amount of beer is drunk on the train,' he told me, 'but it's not like drinking five pints of generic product; it's about having new tastes and new experiences.'

I suggested that there was a learned element to real ale connoisseurship, which makes it akin to railway enthusiasm.

Alan agreed. 'Yes, they both have the historical interest, and you have that completism – people who want to see every type of a certain engine, and people who want to drink every beer by a certain brewery.'

I noticed that at every station stop, the Pig & Whistle would briefly empty before filling up again. 'Where's everyone going?' I asked one of the barmen.

'To the toilet,' he said frankly. (The train toilets are locked on the Epping–Ongar Ale Trains. This is because they are what is picturesquely known as the 'dump type', depositing straight onto the tracks, and there'd be an inundation if they were used on Ale Nights.)

When I bought my next pint, I thought I'd better make a note of the beer. It was called *Cowcatcher*, was brewed by the East London Brewing Company, and it had a specific gravity of 4.8. So it was strong, which is probably why, the next time we called at Ongar, I found it quite exhilarating to be peeing in the roofless Gents under the increasing downpour.

Returning to the train, I bought a third pint and collared Alan again for what was (on my side) a fairly incoherent speculation about railway licensing laws. Alan said he'd summarise them in an email, and we'll come to that in a minute. I 'bailed', as the diesel fanatics say, after three-and-a-half pints. From North Weald, I rode the same double-decker back to Epping, but now there was a different conductor, albeit of a similar vintage to the earlier one. Whereas the previous driver had been a man, this one was a woman. 'She's a Stagecoach driver out of Bow,' the conductor assured me. Being drunk, I complimented him on his vintage, summer-weight, cream conductor's jacket, hoping (I suppose) to earn a 'Mind how you go' when we reached the terminus. It turned out he favoured the equally time-honoured, if less personalised, 'All change!' and I was perfectly happy with that. In fact, I was perfectly happy full stop.

Alan's email arrived a few days after my visit to the Epping–Ongar. It was headlined, 'You don't need a licence for a moving bar,' and summarised government advice relating to the Licensing Act of 2003:

> Railway vehicles and aircraft engaged on journeys are exempted from the requirement to have an authorisation to carry on licensable activities (although a magistrates' court can make an order to prohibit the sale of alcohol on a railway vehicle if this is appropriate to prevent disorder).

So a morbid plot premise of *The Titfield Thunderbolt* – namely that a wealthy alcoholic would fund the reopening of the line because he would be able to drink in the buffet car of the train before the pubs opened – was evidently correct. Alan's email continued ...

> A moving train does not make for an ideal cellar, since the casks will be frequently disturbed, leading to murky-looking pints. To avoid this, it is common practice to serve 'bright' beer, which has gone through the secondary fermentation process in a cask in some other place, usually a brewery cold store. As no extraneous carbon dioxide is introduced during the process of making the beer bright this can still be counted as real ale.

'Real Ale and rail travel somehow go together,' began an article in the April 2018 edition of the *Railway Magazine* entitled 'Forty Years of Real Ale Rambles', 'especially with heritage railways, which often hold beer festivals at stations or on board trains. Both interests appeal to people who love tradition and want to support it in the face of an increasingly corporate world.' The article then explained the source of the tradition.

Real Ale trains began as 'Real Ale Rambles', pioneered by Gerald Daniels, a BR employee and CAMRA member who worked in the area manager's office at Surbiton. Mr Daniels, who wrote a book about closed-down stations called *Passengers No More*, began organising tours using spare carriages hauled by a Class 33 diesel. The first Ramble took place on 1 October 1977; it was advertised in CAMRA's *What's Brewing* magazine and was a sell-out.

The train departed Waterloo at 08.34 calling at Surbiton, Basingstoke, then Bath and Oxford. The 600 passengers visited CAMRA-approved pubs near the stations. 'We reckon the passengers got through about 3,000 pints,' Mr Daniels recalled.

Other Rambles quickly followed. The author of the *Railway Magazine* article, Dave Richardson, had reported on the third tour: a jaunt from Euston to Macclesfield, Stockport and Manchester taking place on 15 April 1978. Of the 400 revellers on the train, he blithely wrote, 'Although they had each drunk ten pints, the free Lancashire hotpot served at the Stockport soaked it all up.' Mr Daniels continued to organise Real Ale Rambles, but in the early 1990s the numbers dried up. Privatisation made the excursions harder to organise, and the success of CAMRA meant that people didn't have to board a special train to access real ale pubs. Then again, the success of the preserved railways means that real ale drinking *on* trains is now a popular activity, and Mr Daniels, who turned eighty in 2017, still operates Real Ale rambles through his company Cookham Travel, although these are 'more sedate affairs than in the early days, when some people really did get "wrecked".'

7

Some Possible Futures

The Gloucestershire Warwickshire Railway

I had come to the Gloucestershire Warwickshire to experience a line that seems exemplary in its forward planning, and one that looks to have the brightest of futures. It bills itself, like several others, as 'the Friendly Line',* and it *is* friendly, which is just as well, because otherwise it might be smug. The Gloucestershire Warwickshire tends to do things right, and those things will be followed up by a big headline in the *Railway Magazine*. In particular, it has been responsible for the most trumpeted line extension to have occurred during the time I was writing this book, namely its move, completed on Good Friday, 30 March 2018, on Broadway, that Cotswold town and – happily for the GW – major tourist destination. Visitor numbers over that Easter week of the reopening were 6,000, double the figure for the same week in 2017.

Broadway is the new northern terminus of the line, but I started my trip along the railway one overcast afternoon at its southern terminus, Cheltenham Racecourse.

---

*The designation 'Friendly Line' might become tyrannical. I foresee a time when all the preserved lines feel obliged to use it, for fear of being asked, 'So you're not friendly, then?'

This station stopped receiving race-day specials from the national network in 1976. Today the Gloucestershire Warwickshire runs its own 'hugely successful' (I quote from the official guide book) race-day specials: champagne is served on the train, and punters are taken to a 'special area' of the racecourse. The GW is the only preserved line able to run race-day specials, which have always been amongst the raciest in the wider sense of the word.

(My dad was both a railwayman and a keen gambler. He liked to boast that every weekday of his life, aside from when he was ill or abroad, he'd backed a horse. Being the possessor of a 'priv' ticket, he was often aboard race-day specials, and he had many stories of his rackety fellow passengers. As a young man, coming back from Ascot, he was joined in his compartment by two well-dressed, gangsterish men carrying attaché cases. They each lit cigars (it was a non-smoker), then one of them took off his jacket, revealing an expensive silk lining. He spread it over a seat, and the two of them began unloading fist-fuls of bank notes from the cases onto it. They had by no means completed this process when one of them glanced towards my father, apparently noticing him for the first time. 'You,' he said, 'out!' My father complied; it would clearly have been dangerous to do otherwise.)

Cheltenham Racecourse Station was re-opened by Princess Anne in 2003. The rebuild of the station secured the GW the Heritage Railway Association Award for Excellence in Overcoming Adversity, because there once *was* adversity hereabouts. The infrastructure of the 21-mile Honeybourne Line (Cheltenham to Honeybourne) was crumbling when BR closed the route in 1976; even so, the preservationists could not afford to buy it. In 1979, BR lifted the track and demolished the stations, thereby – absurdly – making the line affordable to the preservationists, who acquired it in 1981 and began running trains in 1984. The line is today 14 miles long, and every mile of track was relaid by the GW. In *For the Love of Trains*, Denis Dunstone writes that the railway is characterised by 'good PR and great physical effort' – two things that don't often go together.

As part of a preserved line, Cheltenham Racecourse Station is busier today than it was pre-preservation, when it was only ever open on race days. It is also better kept than ever. It has one long, very neat, gravelled platform, bordered with flower beds, and shadowed by pine trees, every one of which seems to harbour a bird box. Pine trees are a specialism of the Gloucestershire Warwickshire. One volunteer explained to me as the train pulled away from Cheltenham Racecourse, 'They don't shed leaves, they look good and they act as a wind shield; and the sound of the wind blowing through them reminds people of the sound of an approaching steam train – and people find that reassuring.'

The old Honeybourne Line formed the main GWR link between the Midlands and the South-West. The line was double-tracked, and handled express trains, including *The Cornishman* (Wolverhampton to Penzance). *The Cornishman* was in part a holiday train, but the Honeybourne Line itself ran through a pretty desirable destination, namely the Cotswolds. Being located in moneyed countryside has been useful to the Gloucestershire Warwickshire: they have rather a good class of volunteer. As my informant told me, 'We have a whole range of professionals: doctors, a lot of skilled engineers, a merchant navy captain ...' The line has a *lot* of active volunteers as well – about 950 – and a high proportion of volunteers to paid staff, of whom the line has only a handful, as against their regional rival, the Severn Valley, which has a hundred. My informant acknowledged the value of a prosperous locality, but added, 'We have people from all over. One of our guards lives on the Isle of Wight. He comes up for one weekend every month; he has breakfast on the ferry on Saturday morning, and dinner going back on it on Sunday night.' He suggested that when it came to preserved railways, once a volunteer always a volunteer. 'You'll find that anyone who has anything to do with steam ... it's addictive.'

Volunteering is a very British thing. The level of volunteering in Britain is high and more or less constant, although the National Council for Voluntary Organisations suspects there has been 'a very slight decline in recent years, possibly explained by people having more calls on their time through multi-media and the Internet.' According to the NCVO, one in four people volunteers once a month, and preserved railways – with their 22,000 volunteers – are the biggest 'employers' of volunteers in the leisure and tourist sector. But whether the Gloucestershire Warwickshire's sanguinity in this respect will endure for long is another matter. On 4 February 2016 an article about preserved railways by Stephen McGrath appeared in the *New Statesman*: 'The majority of ... volunteers are north of fifty years old – "males of a certain age", as one industry worker put it – who would've likely grown up with steam trains in the not so distant past. That means there's a potential demographic time-bomb looming.'

It might be possible to keep recruiting fifty-year-olds, but there's no guarantee of easy pickings even amongst that cohort. The fifty-year-olds of today post-date 'Generation Steam', and so might not be susceptible to railway romance. The first preserved railways were started by young people, often students. The different situation of today implies decadence. I ran the phrase 'demographic timebomb' past the seventy-two-year-old publicity officer of a West Country railway: 'It's very real,' he said. 'Some railways have youth clubs; we don't, but we have endless recruitment campaigns. I think it's like the Army's recruitment crisis. Even if people in their twenties are attracted to the glamour of a steam engine, they find the work very physically demanding. People in their twenties are not so robust as we were. When we were kids, we didn't sit around all day watching TV.'

The *New Statesman* article also discussed the danger of the preserved lines losing old engineering skills. 'Really the big issue in skills shortage is boiler smiths,' Kieran Hands, a (young) heritage railway engineer, was quoted saying: 'because that's a very, very special skill, and every steam loco has a boiler. When a rebuild becomes absolutely massive – when you start talking about hundreds of thousands of pounds – it's normally when the boilers are shot, because it's not just normal welding; it's copper welding. Most people can learn to weld steel; but copper welding is a very specialist skill.'

In 2019, the *Railway Magazine* ran a series of features about the Heritage Skills Training Academy set up by the Severn Valley Railway in 2013 and funded by the Severn Valley Railway Charitable Trust with sponsorship – as of 2019 – from the *Railway Magazine* itself, which would have many blank pages if preserved railways started closing. In the February issue, a sixteen-year-old trainee was interviewed about working on a Hall-class engine that had not steamed since 1986: 'Last week, I was one of the team fitting rivets to the loco frame. We heated them up using propane and oxygen, so they were sweating, and dropped them into the holes. You use a rivet gun to shape the rivet on top, and a jammer to hold it in place underneath.'

The Ffestiniog and Welsh Highland Railways also run a Heritage Skills Training Programme. The slogan is, 'Make the past your future.'

For now, the low wage bill on the Gloucestershire Warwickshire means that fares (and the generous bequests that the GW-type of volunteer tends to leave) can be spent directly on improving the service and the line: 'Our permanent way is second to none,' my informant told me. 'We use a lot of concrete. Laverton to Broadway is welded track.' Our conversation was occurring, by the way, in a well-maintained BR Mark 1 buffet generously stocked with Return to Broadway Celebration Ale.

Even before reaching Broadway, the line was successful. 'We have more visitors in each successive year. Last year, we had 101,000 fare-paying passengers.' We were passing over the Stanway Viaduct (fifteen arches, 42 feet high) which, along with the Greet Tunnel – the second longest on a preserved railway – gives a grandeur to the line. 'On the North York Moors Railway,' said my informant, 'because you're in a valley, you're often running through cuttings, whereas we have *scenery*.'

Broadway Station was completely rebuilt from the

rubble of the old one. Some bits of the rubble were useful: new ticket barrier supports were cast using the originals – found during the first excavation of the site – as models. The new station is exactly the same as the old one (built in 1904). If you compare photographs of the station prior to its demolition in 1963 (thirteen years before the closure of the line) with those taken since its rebuilding, the only difference seems to be that the latter are in colour. The rebuilt station is a bit disconcerting: brand new, yet old-looking. How often is a modern-day building contractor required to create a Victorian fire-place, as in the Broadway waiting room? It was of course in the Great Western colours of light and dark stone, and I could still smell the paint, because the contractors hadn't quite departed: pop music seeped from a builder's radio.

Incidentally, Broadway is in Worcestershire, further justifying the line's use of the familiar acronym GWR, so here is the third GWR that the writer on railways must contend with: the others being the historic one and the current national network operator serving the West Country.

The Broadway station-master smiled and nodded at me as I admired his station. He had a flower in his buttonhole. My informant informed me that the railway – let's call it the GWR now – would be 'consolidating' for a while before pushing on to Honeybourne and a connection with the national network. 'Network Rail have already prepared a platform for us.'

The East Lancashire Railway

I felt like a Lowry Man as I approached Bury Bolton Street Station, walking across the car park that was the old goods yard, with its high, black, arcaded walls. Black smoke hung over the station, which is right in the middle of town: a Standard Class 4 Tank engine was fuming away inside.

In decades past, there would have been many other sources of smoke and steam hereabouts, but the textile industry of Bury died in the 1960s, as the many truncated mill chimneys of the district attest. Dr Beeching smelt blood. I quote from the *Official Guide* to the East Lancashire Railway: 'the areas around Bury and the Rossendale Valley were hit hard by these proposals, with only the Bury to Manchester service (now part of Metrolink) surviving the closure threat.'

The East Lancs runs through fields and parkland for much of its route, but it's the most urban of the big preserved railways. Its HQ is bang in the middle of Bury, and we can think of Bury Bolton Street as the crux of a softened capital letter L. The line goes north from Bury Bolton Street to Rawtenstall, along a section of line closed to passengers in 1972 and to coal trains from 1980. The East Lancashire Railway Society – formed in 1968 – reopened this between 1987 and 1991. The line also heads east from Bury Bolton Street to Heywood, along a section closed in 1970 and re-opened in 2003; and there are plans to go further east – to Castleton. At the moment, the line is 12 miles long.

Bury Bolton Street is a bruised-looking station of red brick and BR Midland Region maroon. It's half 1950s and half Victorian. Pre-Grouping it was operated by the

Lancashire & Yorkshire Railway, that most industrial of railways. There's a compelling anti-glamour about the East Lancs, which has more diesel engines than any other preserved railway. But this was a steam day, and I hadn't come to Bury for the engines. I'd come to make a point about the future of preserved railways ...

The East Lancs looked in good shape for the future, by which I mean the train on which I rode had a balance of young and old. To start with the old ... On boarding the train at Bury Bolton Street, I found myself sitting opposite an elderly traditionalist with a pile of railway magazines in front of him. Indicating the car park, he said to his wife, who seemed to be trying to go to sleep, 'That used to be the goods yard, before cars came along and messed everything up.' He then began discussing the revived fortunes of Salford with the equally venerable buffet car attendant.

'You wouldn't give tuppence for Salford in the 1950s,' said the attendant. 'There was no reason to go there.'

'You probably wouldn't come out if you *did* go there,' said the traditionalist.

'Now,' said the attendant with incredulity, 'it's *fashionable*.'

In the carriage behind ours, 'Happy Birthday' was being sung. I listened for the name: 'Happy birthday, dear ... Athena.' Going by the name, I didn't think this would be a *teenage* birthday, and, craning to look down the aisle, I saw two heart-shaped birthday balloons tethered to a seat: a number eight and a zero. But this was being counterbalanced, demographically, by the performance from the carriage in front: a party of schoolchildren singing a song about stripy socks ... which dissolved into screams of delight as we ran through Bury North Tunnel, as if this were a ghost train.

The East Lancs makes a big effort to attract school parties, and to reach out beyond Generation Steam. It has a Learning and Audience Development Manager, Matthew Britten, who believes that preserved railways *are* facing a demographic crisis, and he is researching this in the hope of writing a PhD on the subject, while at the same time trying to avert that crisis at the East Lancs. For the moment, the number of volunteers on preserved lines is holding steady, but Britten believes the preserved lines should take note that the National Trust reported a drop in volunteer numbers in 2018, citing an ageing population, longer working hours and a decline in disposable income – and the decline was steepest among young volunteers.

In 2019, Britten inaugurated an industry placement scheme for seven sixteen-to-eighteen-year-olds. They were involved in 'retail, meeting-and-greeting and general event support', and all of these students have since become regular volunteers on the line. The East Lancs has offered apprenticeships in Administration and Engineering for a number of years. The railway has a primary education programme that brings 3,000 children a year to the railway. It has partnerships with seven local secondary schools, and a secondary education programme teaching STEM (science, technology, engineering and mathematics) subjects is in the offing. The East Lancs also has a Kids' Club, presided over by Buffers the cat. (Members receive a regular newsletter and a card entitling them to discounts on the railway.)

The East Lancs is firmly rooted in its locality. This is densely populated, which helps explain why a high proportion of the ELR's 4,500 members are active volunteers: about 750. 'That's the thing about our line,' the train guard told me as he checked my ticket: 'Bury, Ramsbottom, Rawtenstall, they're all *towns*, and the railway has done a lot to help them.' It might, in fact, have done too much, which has given people ideas.

Before coming to Bury I'd spoken to Bruce Williamson, press officer for Railfuture, an advocacy group for passengers and railway freight hauliers. 'The preserved lines are great in their own right for all the employment they create, and the knock-on effects that are well known,' he said. 'But something else that's interesting about them ... they preserve the track bed, and that might eventually be used to re-open a conventional railway.' He conceded that most preserved lines would resist the idea – and this is where the East Lancs comes in.

As we rumbled towards Rawtenstall, I opened the latest copy of the *ELR Review*. First up was 'News From the President': 'Following press speculation, I know many of our members have been really worried about the unbelievable proposals which the Rossendale Council were putting forward to introduce commuter trains along the valley and in effect undermine our heritage railway operation.'

The plan – for trains to come south along the East Lancs line to Bury Bolton Street, from where passengers would walk to the Bury Interchange of the electrified Manchester Metrolink – would seem to be incompatible with the operation of steam and heritage diesel trains. At the time of writing, it seems unlikely to come to fruition.

Back to the *ELR Review*: 'We should all be pleased and relieved to note that no proposals for commuter trains along the valley appear in the document.' The ELR people don't believe there's a business case for the proposed commuter service, and they point to the benefits their own line brings to the local communities (a hundred jobs) and the tourist income of the North-West. But Rossendale Council hasn't given up on the scheme, and the President was forced to address another threat to the railway, this one to its plans for an extension to Castleton: 'We do need to keep an eye on the Transport for Greater Manchester idea for a possible Tram Train development from Heywood to Rochdale which was hinted at in the TFGM 2040 strategy. Again, there are no firm proposals, so we should simply get on with what we do best and actively support our award-winning ELR.'

Other preserved lines are being eyed by the bigger players of the national network. After my visit to the West Somerset Railway, I read that the Great Western Railway (the modern one) is to begin an 'experimental' diesel shuttle service from Taunton to the West Somerset's line, so that Minehead, for instance, will be reconnected to the national network. The WSR welcomes this: 'We are proud to be running our heritage steam trains alongside their Great Western successors.' But you can see how tension might develop. The Bodmin & Wenford Railway recently received a grant from the GWR to improve facilities at Bodmin General. The GWR aspires to run a diesel commuter service from Bodmin General to connect with the main line at Bodmin Parkway. The Bodmin & Wenford welcomes this, 'as long as it doesn't interfere with our core business.'

In a moment of devilment, I suggested to a B&W spokesperson that their core business might be called, by some, merely playing with trains. Why should this be allowed to stand in the way of a new commuter service, with all its social and environmental benefits?

The spokesperson, retaining an equable tone, said, 'Our core business is running a steam railway for tourists. And another thing – it's *our* line.'

On the East Lancs line, we had stopped at Ramsbottom Station en route to Rawtenstall. 'Tesco Metro,' the elderly traditionalist muttered with distaste (there is one next to the station.) He then turned his attention to the young man working on the platform. 'And look at that platform guard,' he said. 'He's wearing just any old top. I'd tell him, "You're improperly dressed, young man; you should be wearing a red tie."' He seemed to be addressing his wife, who was at least half asleep. Not for long, however. 'Wake up, love,' he said, nudging her arm. 'We're coming to a level crossing.'

Southwold

The Southwold Railway, which connected that salubrious Suffolk port to Halesworth 9 miles inland, closed in 1929, killed by a new bus service. Gaumont News added insult to injury by making a satirical film about the last train, headlined 'The Railway that is a Real Joke', with sarcastic captions such as 'Has been known to complete the journey of 9 miles to Southwold in fifty minutes!'

Even though the line paid a dividend in some years, and was well-maintained and run, and even though it had carried the one million bricks used in the building of the Southwold lighthouse, it had always been the butt of jokes: something to do with the unusual 3-foot gauge; eccentric-looking engines given a Far Tottering and Oyster Creek aspect by disproportionately high chimneys; the makeshift combination, on a typical train, of passengers and goods (usually fish) ... and many of the passengers in summer were holidaymakers, hence a certain levity. A local cartoonist, Reg Carter, produced a series of humorous postcards in Rowland Emett-like style depicting misadventures on the line.

In 1996 the Southwold Railway Society, which is now the Southwold Railway Trust, was formed to rebuild and re-open the line. As a regular summer visitor to Southwold, I became a member, and so the recipient of the monthly SRT newsletter. There might be a report on, say, the members' annual summer ramble along the route of the line: 'We reached the old bridge near the harbour: the rails are sinking into the mud ... More brambles than ever this time.'

Every edition featured a picture of John Bennett, chairman of the SRT, wearing a broad-brimmed hat and a sort of long dustcoat while standing at the controls of some strange railway vehicle. I had recognised, in my certificate of SRT membership, the dandified style of John Bennett: it was written in copper plate, with a fancy border. John is an architect, and any missive from him usually includes beautifully drawn little plans and maps, always with 'North' decoratively depicted. His drawings of the old trackbed feature landmarks like 'The Heronry' (cue drawing of a heron), Bird's Folly, Angel Marsh. But Bennett combines an apparent whimsy with determination and technical skill. He created in Southwold the

Electric Picture Palace cinema, to 'recapture the experience of cinema-going in the mid-twentieth century', so, although it's tiny with only seventy seats, the EPP has a balcony, a uniformed commissionaire and intermissions in which – to the profound shock of first-time visitors – an organist rises from beneath the stage.

The aim of the Southwold Railway Trust is to gradually buy up the land over which the old Southwold Railway ran. Meanwhile, they have built and opened a one-acre railway theme park on the site of an old gasworks near the harbour where one branch of the line used to terminate. I went there on a hot Saturday afternoon in 2018, soon after it had opened. John Bennett was not around, but I recognised his handiwork in the pretty blue banner hanging over the entrance gate reading 'Steamworks'; also in the landscaped pond (which Mr Bennett calls a lake) around which a 7¼-inch miniature railway – the Blyth Valley Light Railway – has been constructed.

Steamworks was apparently doing quite well, but this was a quiet time. Two volunteers were present (cheerful men in their seventies or so). I enquired about a ride on the Blyth Valley Light Railway. They looked at each other, came to a resolution. That would be fine, except that the regular locomotive – a hybrid petrol-electric with a transmission from an electric wheelchair – was out of action, so the motive power would be an adapted lawnmower engine. They also explained that any miniature train required both a driver and a guard, so I would be outnumbered by crew on the ride. Even so, the Bennett-ish formalities were observed, and I was sold, for one pound, an Edmondson-style ticket. As we set off around the lake/pond, I talked to the guard who, gesturing at the site with his green flag, said: 'This'll all be landscaped, of course.' He was indicating a loco shed; a pavilion-like station; a short length of 3-foot gauge track along which the SRT hopes to run its replica of a Southwold Railway engine (when they have the money to build the replica, that is). They also have a Belgian tramcar of the 3-foot gauge.

But Halesworth is the ultimate goal. 'We've already bought a field near Wenhaston,' said the guard, 'to stop the antis getting it.' There are a significant number of antis – people who don't want the line re-created.

I asked the guard what their objection was.

'They say it disturbs the wildlife, and they don't want the visitors.'

Without a railway, Southwold has become an exclusive resort – 'Hampstead on sea,' it's known as. Perhaps the antis fret that, if the railway returned, it would be a bit less Hampstead and a bit more Lowestoft.

As we completed our final circuit of ... well, let's say 'the water', I suggested to the guard of the miniature train that it would take a heck of a long time to acquire all the trackbed between Halesworth and Southwold, and it wouldn't be easy to fight off the antis.

'If the Middy can do it, we can,' he said. (The Middy was once the Mid-Suffolk Light Railway, an agricultural line built under the 1896 Act and closed in 1952. A half-mile stretch of the same line has been preserved under the same name by the Mid-Suffolk Light Railway Society since 1992.) 'Of course,' the Southwold man continued, 'it won't happen in our lifetimes,' and he was clearly including me in this, which was a bit chastening. Then again, this is the fascinating thing about preserved railways: the meddling with time. What this man was saying was that, many years hence the Southwold Railway Trust might succeed in going many years back into the past.

A year later, late on another hot Saturday afternoon, I was back at Steamworks. The project had moved on, as Mr Bennett's schemes tend to do. Wearing a very nice white linen suit, he and the Southwold vicar – a rail enthusiast who used to work for BR – were sitting outside the station on deckchairs. The pond/lake was beginning to bed in, and Mr Bennett would have me know that it had been officially classified as a lake on the Ordnance Survey map. A train on the miniature-gauge Blyth Valley Light Railway – with a dozen passengers on board – was just departing from a newly built halt, surmounted by a pediment in which a clock was mounted. Under the station canopy a wind-up gramophone played 1920s jazz. The lawnmower engine was nowhere to be seen: today's motive power was a pretty little side-tank which burned, I was told, 'Anthracite beans' meant for barbecues. Apparently, Steamworks had taken a hundred pounds that day on train rides alone.

Mr Bennett was about to head back into town, probably – this being a Saturday – to put on the dinner jacket in which he presides at the cinema. I walked with him, past the golf course and onto Southwold High Street, where he lives. You can tell Mr Bennett's house because he's constructed a sort of turret-come-viewing platform on the roof, so he can see the sea.

There had been no new acquisitions of trackbed between Southwold and Halesworth since I'd last visited and, given the long-term nature of the Southwold

Railway Trust's project, something was bothering me: an article I'd read in that month's *Railway Magazine* (August 2019). 'Steam lines face double threat in crackdown on coal-burning. Britain's preserved railways are fighting for their future in the face of a Government crackdown on carbon emissions.'

The crackdown was the result of the government's commitment to net zero carbon emissions by 2050: 'Domestic coal- and wood-burning in homes will be restricted by 2022, coal-fired electricity generation will end by 2025, and diesel-only trains could be phased out by 2040.' The Department of the Environment had indicated that it was not the intention to prevent the burning of coal in steam engines and other historic and decorative uses of coal, but there was no absolute guarantee of this. The preservationists were also worried that coal supplies would dry up, and they feared targeting by protestors. Steve Oates of the Heritage Railway Association was quoted as providing the counter arguments, touching on all the many social and economic benefits of the preserved lines, and the environmental ones: 'They provide green corridors, on which flora and fauna are protected ... They help keep holidaymakers in the UK rather than flying off on polluting jets ...'

Trepidatiously, I mentioned some of this to Mr Bennett, and I was glad to see that he wasn't in the least fazed. 'Oh, I know our railway will have to be carbon-neutral,' he said. 'That's why I'm looking at the possibility of hydrogen locomotives.' He then said some technical things I couldn't follow, but which concluded, ' ... and you'll still be able to have steam coming out of the chimney.'

That, I think, is the main thing.

Select Bibliography

Chris Arnot, *Small Island by Little Train: A Narrow-Gauge Adven*ture, Basingstoke, AA Publishing, 2017

Jonathan Brown, *The Railway Preservation Revolution: A History of Britain's Heritage Railways*, Barnsley, Pen & Sword, 2017

Ian Carter, *Railways and Culture in Britain: the Epitome of Modernity*, Manchester, Manchester University Press, 2001

Ian Carter, *British Railway Enthusiasm*, Manchester University Press, 2014

Denis Dunstone, *For the Love of Trains: The Story of British Tram and Railway Preservation*, Shepperton, Ian Allan, 2007

Bob Gwynne, *Railway Preservation in Britain*, London, Shire, 2011

John Hadfield, *Love on a Branch Line*, London, Hutchinson, 1959

Simon Jenkins, *Britain's 100 Best Railway Stations*, London, Viking, 2017

Ian Marchant, *Parallel Lines: or, Journeys on the Railway of Dreams*, London, Bloomsbury, 2003

L. T. C. Rolt, *Narrow Boat*, London, Eyre & Spottiswoode, 1944

L. T. C. Rolt, *Railway Adventure*, London, Constable, 1953

Brian Sibley, *The Thomas The Tank Engine Man: The Life of Reverend W. Awdry*, Oxford, Lion Publishing, 2015
Paul Smith and Keith Turner, *Railway Atlas Then and Now*, Shepperton, Ian Allan, 2012
J. B. Snell, *Jennie*, London, Thomas Nelson and Sons, 1958
Nicholas Whittaker, *Platform Souls: The Trainspotter as 20th-Century Hero*, London, Gollancz, 1995
Michael Williams, *The Trains Now Departed: Sixteen Excursions into the Lost Delights of Britain's Railways*, London, Preface, 2015
John Winton, *The Little Wonder: 150 Years of the Festiniog Railway*, London, Michael Joseph, 1986

Picture Credits

1 E2 class locomotive. From New Build Steam website.
2 Dr Richard Beeching, Chairman of British Railways, reopens the Dart Valley Railway, South Devon Railway, 21st May 1969. Photo by Brennan/Mirrorpix/Getty Images
3 Tim Rolt, founder of the preserved railway movement. Photograph from the Rolt family archive.
5 Tornado departing Redmire station, with Bolton Castle on the skyline. Photo by Maurice Burns.
8 Garratt steam locomotive on the Welsh Highland Railway in North Wales, photographed at Porthmadog. Mike Rex/Alamy Stock Photo

All other photographs are the author's own.

While every effort has been made to contact copyright-holders of illustrations, the author and publishers would be grateful for information about any illustrations where they have been unable to trace them, and would be glad to make amendments in further editions.